Innovation and Research Policies

NEW HORIZONS IN THE ECONOMICS OF INNOVATION

General Editor: Christopher Freeman, *Emeritus Professor of Science Policy, SPRU – Science and Technology Policy Research, University of Sussex, UK*

Technical innovation is vital to the competitive performance of firms and of nations and for the sustained growth of the world economy. The economics of innovation is an area that has expanded dramatically in recent years and this major series, edited by one of the most distinguished scholars in the field, contributes to the debate and advances in research in this most important area.

The main emphasis is on the development and application of new ideas. The series provides a forum for original research in technology, innovation systems and management, industrial organization, technological collaboration, knowledge and innovation, research and development, evolutionary theory and industrial strategy. International in its approach, the series includes some of the best theoretical and empirical work from both well-established researchers and the new generation of scholars.

Titles in the series include:

Foundations of the Economics of Innovation
Theory, Measurement and Practice
Hariolf Grupp

Industrial Organisation and Innovation
An International Study of the Software Industry
Salvatore Torrisi

The Theory of Innovation
Entrepreneurs, Technology and Strategy
Jon Sundbo

The Emergence and Growth of Biotechnology
Experiences in Industrialised and Developing Countries
Rohini Acharya

Knowledge and Investment
The Sources of Innovation in Industry
Rinaldo Evangelista

Learning and Innovation in Economic Development
Linsu Kim

The Economics of Knowledge Production
Funding and the Structure of University Research
Aldo Geuna

Innovation and Research Policies
An International Comparative Analysis
Paul Diederen, Paul Stoneman, Otto Toivanen and Arjan Wolters

Innovation and Research Policies

An International Comparative Analysis

Paul Diederen

Senior Researcher, Agricultural Economics Research Institute (LEI), The Hague, The Netherlands

Paul Stoneman

Research Professor, Warwick Business School, University of Warwick, UK

Otto Toivanen

Assistant Professor of Economics, Helsinki School of Economics, Finland

Arjan Wolters

Researcher, Agricultural Economics Research Institute (LEI), The Hague, The Netherlands

NEW HORIZONS IN THE ECONOMICS OF INNOVATION

Edward Elgar
Cheltenham, UK • Northampton, MA, USA

Published by
Edward Elgar Publishing Limited
Glensanda House
Montpellier Parade
Cheltenham
Glos GL50 1UA
UK

Edward Elgar Publishing, Inc.
136 West Street
Suite 202
Northampton
Massachusetts 01060
USA

A catalogue record for this book
is available from the British Library

Library of Congress Cataloguing in Publication Data
Innovation and research policies : an international comparative
    analysis / Paul Diederen ... [et al.]
        p.   cm.
    Includes bibliographical references.
    1. Technological innovations.   2. Technological innovations–
Economic aspects.   3. Research, Industrial.   4. Competition,
International.   I. Diederen, Paul, 1959–.
HD45.I53718   2000
338'.064–dc21   99-34898
                                                                    CIP

ISBN 1 84064 027 8

Typeset by Manton Typesetters, Louth, Lincolnshire, UK.
Printed and bound in Great Britain by
Creative Print and Design (Wales), Ebbw Vale

Contents

Figures

Tables

Abbreviations

Various organizations

CEC	Commission of the European Communities
COREPER	Comité Representative Permanente
CERN	European Laboratory for Particle Physics
CERT	Committee for Energy, Research and Technology
CREST	Comité de la Recherche Scientifique et Technologique
EBRD	European Bank of Reconstruction and Development
EMU	European Monetary Union
EP	European Parliament
EuroPES	European Public Expenditure System
ESA	European Space Agency
IRDAC	International Research and Development Advisory Committee
LEI	Landbouw-Economisch Instituut (Agriculture Economics Research Institute)
MERIT	Maastricht Economic Research Institute on Innovation and Technology
NASDAQ	National Association of Security Dealers Automated Quotations
OECD	Organization for Economic Co-operation and Development

United Kingdom (institutions and programmes)

CBI	Confederation of British Industry
CST	Council for Science and Technology
DfEE	Department for Education and Employment
DERA	Defence Evaluation and Research Agency
DOE	Department of the Environment (now: DETR, including Transport and Regions)
DOT	Department of Transport (now: DETR, including Environment and Regions)
DSAC	Defence Scientific Advisory Council
DTI	Department of Trade and Industry
DUTC	Dual Use Technology Centre
HEFC	Higher Education Funding Council
ITC	Innovation and Technology Counsellors

MAFF	Ministry of Agriculture, Fisheries and Forestry
MOD	Ministry of Defence
NHS	National Health Service
OST	Office of Science and Technology
OSTEMS	Overseas Science and Technology Expert Missions
OTIS	Overseas Technical Information Service
PTP	Postgraduate Teaching Partnership
SIUS	Support for Industrial Units Scheme
TCS	Teaching Company Scheme

The Netherlands (institutions and programmes)

AWT	Adviesraad voor Wetenschap- en Technologiebeleid (Advisory Council for Science and Technology Policy)
BTS	Bedrijfsgerichte Technologische Samenwerkingsprojecten (Company oriented Technological Co-operative Projects)
CBS	Centraal Bureau voor de Statistiek (Central Statistical Office)
CoCo	Coördinerend Comité voor Europese Integratie en Associatie (Co-ordination Committee for European Integration and Association)
CPB	Centraal Planbureau (Central Planning Bureau)
DLO	Dienst Landbouwkundig Onderzoek (Agricultural Research Service)
ECN	Energieonderzoek Centrum Nederland (Dutch Center for Energy Research)
EET	Programma voor Economie, Ecologie en Technologie (programme on Economy, Ecology and Technology)
EZ	Economische Zaken (Ministry of Economic Affairs)
GTI's	Grote Technologische Instituten (the large public research institutes)
ICES-KIS	Interdepartementale Commissie Economische Structuurversterking-KennisInfraStructuur (Interministerial Committee on Reinforcing Economic Structures-Knowledge Infrastructure)
IMK	Instituut voor Midden- en Kleinbedrijf (Institute for SMEs)
IOP	Innovatiegerichte Onderzoeksprogramma's (Programmes Oriented towards Innovation)
KB	Koninklijke Bibliotheek (Royal Library)
KIM	Kennisdragers in het Midden- en Kleinbedrijf ('Knowledge-carriers' in SMEs)
KNAW	Koninklijke Nederlandse Academie voor Wetenschappen (Royal Dutch Academy of Sciences)
LNV	Landbouw. Natuurbeheer en Visserij (Ministry of Agriculture, Nature Conservation and Fisheries)

LTO	Land- en Tuinbouw Organisation (Confederation of Dutch Agricultural and Horticultural Enterprises)
LUW	Landbouw Universiteit Wageningen (Agricultural University Wageningen)
NRLO	Nationale Raad voor Landbouwkundig Onderzoek (National Council for Agricultural Research)
NWO	Nederlandse organisatie voor Wetenschappelijk Onderzoek (Research Council)
OCV	Overleg Commissie Verkenningen (Consultative Committee on Government Surveys)
OC&W	Onderwijs, Cultuur en Wetenschap (Ministry of Education. Culture and Science)
PBTS	Programmatisch Bedrijsgerichte Technologie Stimulering (Programmes for Business Oriented Technology Stimulus)
RGO	Raad voor GezondheidsOnderzoek (Council for Health Research)
RWTI	Ministeriële Raad voor Wetenschap, Technologie en Informatiebeleid (Ministerial Council for Science. Technology and Information Policy)
SER	Sociaal-Economishe Raad (Social-Economic Council)
STW	Stichting Technologie en Wetenschap (Foundation for Technical Sciences)
TNO	Nederlandse organisatie voor Toegepast Natuurwetenschappelijk Onderzoek (Dutch Organization for Applied Scientific Research)
TOK	Technologisch OntwikkelingsKrediet (Technological Development Credit)
TTI	Technologische Top Instituten (Leading Technological Institutes)
V&W	Verkeer en Waterstaat (Ministry of Transport, Public Works and Water Management)
VNO-NCW	Verbond Nederlandse Ondernemingen – Nederlands Christelijk Werkgeversverbond (Confederation of Netherlands Industry and Employers)
VROM	Volkshuisvesting, Ruimtelijke Ordening en Milieu (Ministry of Housing, Spatial Planning and Environment)
VSNU	Vereniging Samenwerkende Nederlandse Universiteiten (Dutch Association of Universities)
WBSO	Wet Bevordering Speur- & Ontwikkelingswerk (Law on the stimulation of R&D)
WRR	Wetenschappelijke Raad voor Regeringsbeleid (Scientific Council for Government Policy)

WVS   Welzijn, Volksgezondheid en Sport (Ministry of Health)

France (institutions and programmes)

ANRS   Agence National pour la Recherche sur le SIDA (National Agency for AIDS Research)

ANVAR   Agence Nationale de Valorisation de la Recherche (National Agency on the Valorisation of Research)

BCRD   Budget Civil de Recherche et de Développement technologique (Civil Budget on Research and Technological Development)

CEA   Commissariat à l'Energie Atomique (Institute for Nuclear Energy)

CIRST   Comité Interministeriel de la Recherche Scientifique et Technique (Interministerial Committee on Scientific and Technological Research)

CIR   Crédit d'Impôt Recherche

CNES   Centre National d'Etudes Spatiales (National Centre on Space Research)

CNRS   Centre National de la Recherche Scientifique (National Centre of Scientific Research)

CRITT   Centres Regionaux d'Innovation et de Transfert de Technologie (Regional Innovation and Technology Transfer Offices)

CRT   Centres de Resources Technologiques (Technological Resource Centre)

DGA   Délégation Générale pour l'Armement ('Agency' for Arms Procurement

ENA   Ecole Nationale d'Administration

EPIC   Etablissement Public à caractère Scientifique et Technologique

EPST   Etablissement Public à caractère Industriel et Commercial

FCPI   Fonds Commun de Placement pour l'Innovation

GPT   Grands Programmes Technologiques (Large Technology Programmes)

INFREMER   Institut Français pour l'Exploitation de la Mer Institute for Marine Research

INRA   Institut National de la Recherche Agronomique (National Research Institute on Agriculture)

INSERM   Institut National de la Santé et de la Recherche Medicale (National Research Institute on Health)

MEFI   Ministère de l'Economie, des Finances et de l'Industrie (Ministry of Economics, Finance and Industry)

MENRT   Ministère de l'Education Nationale de la Recherche et de la Technologie (Ministry on Education, Research and Technology)

Innovation and research policies

ORSTOM   Institut Français de Recherche Scientifique pour le Développe-
          ment et Coopération (Institute for Research on Development
          and Co-operation)
OST   Observatoire des Sciences et des Technologies (Observatory
          on Sciences and Technologies)
PLF   Projet de Loi de Finance

Finland
MTI   Kauppa- ja Teollisuusministeriö (Ministry of Trade and Indus-
          try)
STPC   Valtion Tiede Ja Teknologianeuvosto (Science and Technol-
          ogy Policy Council)
SITRA   Suomen Itsenäisyyden Juhlarahasto (Finnish National Fund
          for R&D)
TEKES   (Teknologian kehittämiskeskus) Technology Development Cen-
          tre
VTT   Valtion Teknillinen Tutkimuskeskus (Technical Research Cen-
          tre)

Germany
AiF   Arbeitsgemeinschaft industrieller Forschungsverein (Commu-
          nity of Industrial Research Organisations)
ATI   Agentur für Techniologietransfer und Innovationsförderung
          (Agency for Technology Transfer and Stimulation of Innova-
          tion)
BLK   Bund-Länder Kommission für Bildungsplanung und Forschungs-
          förderung (Federation-State Committee on Education and
          Research)
BMBF   Bundes Ministerium für Bildung und Forschung (Federal Min-
          istry for Education and Research)
BMWi   Bundes Ministerium für Wirtschaft (Federal Ministry for Eco-
          nomic Affairs)
BMVg   Bundes Ministerium für Verteidigung (Federal Ministry for
          Defence)
DARA   Deutsche Agentur für Raumfahrtangelegenheiten (German
          Agency for Space)
DFG   Deutsche Forschungs Gesellschaft (German Research Soci-
          ety)
DLR   Deutsche Anstalt für Luft- und Raumfahrt (German Institute
          for Space and Aviation)
EFA   Service Elektronischer Förderantrag
FRG   Federal Republic of Germany

GDR	German Democratic Republic ('East-Germany')
HLT	Hessische Landesentwicklungs- und Treuhandgesellschaft
MPG	Max Planck Gesellschaft
RKW	RationalisierungsKuratorium der deutschen Wirtschaft
TTZ	Technologiespezifischen und branchorientierte TransferZentern (Technology Specific and Sectororientated Transfer Centres)

Israel

ARO	Agricultural Research Organization
IRO	Industrial Research Organization
JIIS	Jerusalem Institute for Israel Studies
MATIMOP	Israeli Industry Center for R&D
MIT	Ministry of Industry and Trade
MSA	Ministry of Science and Arts
OCS	Office of the Chief Scientist

Japan

| MITI | Ministry of International Trade and Industry |

Singapore

EDB	Economic Development Board
NSTB	National Science and Technology Board
PSB	Singapore Productivity and Standards Board

Taiwan

ITRI	Industrial Technology Research Institute
MoEA	Ministry of Economic Affairs
NSC	National Science Council

Other

GBOARD	Gross Budget Outlays on Research and Development
GDP	Gross Domestic Product
GERD	Gross Expenditures on Research and Development
HERD	Higher Education Research and Development
ICT	Information and Communication Technology
SEB	Science and Engineering Base (UK)
SET	Science Engineering and Technology (UK)
SMEs	Small and Medium-sized Enterprises
TNC	Transnational Corporation

Acknowledgements

The basis for this book was laid in two studies: the first, a research project, 'Methods and Approaches for the Definition of Research Budgets', carried out by Paul Diederen, Paul Stoneman and Otto Toivanen for Directorate General XII of the Commission of the European Community; the second, a study, 'Monitoring Science and Technology Policy III', conducted by Arjan Wolters and Mark Hendriks for the Ministry of Economic Affairs in the Netherlands. We should like to express our gratitude to everybody who contributed to this book, to the respondents who granted us interviews, to colleagues in various countries who helped us to collect written materials and who read our drafts and made valuable suggestions for amendments. Paul Diederen and Arjan Wolters thank the Agricultural Economics Research Institute (LEI) in The Hague, the Netherlands, for giving us the opportunity to update our reports and transform them into this volume. Otto Toivanen gratefully acknowledges financial support from the Academy of Finland and the Yrjö Jahnsson foundation, and the hospitality of the department of economics of MIT. Only the authors are responsible for the contents of this book. It does not reflect the opinion of the Commission of the European Communities of the European Union or the views of the Dutch Ministry of Economic Affairs.

1. Introduction

The industrialized countries of the OECD invest more than 2 per cent of their gross domestic product in research and development activities (2.15 per cent in 1995). Between 1 and 1.5 per cent of aggregate worktime (measured in full time equivalents) in OECD countries is spent on research and development (R&D). These countries together spend more than $400 billion on R&D annually, which comes down to more than $400 per capita. This investment in formal R&D is complemented by investments in informal R&D (for which no figures are available) and by investments in education and training, which amount to between 5 and 8 per cent of GDP in most industrialized countries. Together these investments help to determine our scope of opportunities in the future.

Governments are heavily involved in this investment process. Just under half of the investment in R&D is financed by governments. Most of the funds flow to universities and public research institutes, but there is also public support for private businesses to carry out R&D activities. About two-thirds of it is performed by the private sector; one-third is done in public sector organizations. In addition, governments usually run the main part of the educational system.

In every developed country, dozens of policy makers, spread over various ministries and participating in numerous committees, are involved in decision making on these activities on a daily basis. They deal with a myriad of research and development programmes and manage countless technology and innovation policies. They are advised and supported by hundreds of staff and stakeholders. They influence or even direct the activities of thousands of researchers. The results of these processes affect the lives of millions. This process of policy making is the topic of this volume.

1.1 ABOUT THIS BOOK

Objectives of the Book

This book brings together the main themes that figure in current policy debates on science and technology in Europe. We look at the policy objectives and issues for R&D policies and related policy fields such as science and education. Main national R&D policy objectives at a general level concern matters such as economic competitiveness, social well-being and national security. However, the way these general objectives are translated into actual policies is country- and time-specific. Different countries face different problems, feel different needs and choose different solutions.

We explore the innovation and technology debate by studying R&D policy making in five European countries. We try to understand the policy issues by considering the debates within their specific national context. In particular, we look at the structural characteristics of the system within which policy is made. We consider the allocation of responsibilities regarding R&D policies, the processes by means of which R&D priorities are set and decisions on budgets are taken at the national level. We deal with the funds allocation mechanisms, the procedures followed and the criteria used.

Research Procedures

In the following chapters we look at R&D policy in five case study countries: the UK, the Netherlands, France, Finland and Germany. The UK, France and Germany were chosen as they are the three largest economies in the European Union. The Netherlands and Finland are smaller countries and were chosen partly to provide a contrast. In addition, in a study such as this, local knowledge is of some value, and as there are Dutch, British and Finnish in our team, there were some advantages in this selection of countries. For comparison a limited amount of supplementary material about a number of non-European countries was gathered. Information on the US, Japan, Israel, Taiwan and Singapore is presented in one chapter.

The five case studies are based upon a series of interviews with senior civil servants involved in the R&D process in each of the countries, supplemented with other information to be found in relevant government publications. It is noteworthy that a lot of the information relevant to the topics of this book can be accessed through the Internet these days. Most ministries in OECD countries dealing with R&D policies maintain webpages that give ample information on their policies. In many cases they also provide up-to-date statistics on R&D performance and policy results.

Our interviewees were primarily located in those national institutions with the greatest involvement in the R&D process. In addition, our own knowledge of existing processes and procedures based upon contact with (and sometimes involvement in) the R&D process has been introduced where appropriate. The interviews were based upon a *pro forma* questionnaire that was used to guide discussions. A copy of this questionnaire is included as an appendix to this chapter. The questionnaire was supplied to interviewees prior to the interview. Each interview had its own dynamic and, as such, different aspects of the questionnaire were stressed in different interviews and also, when appropriate, discussions went outside the strict confines of the questionnaire. We have tried to supplement the purely factual material with personal interpretations of the base material and so the material presented should not necessarily be taken as representing (totally) the views of the interviewees.

It should be noted that even at this early stage, however, any detailed investigation of the process of the setting of R&D budgets, priorities and policies can only be a snapshot taken at a particular point in time. Policies and attitudes change over time, not only with changes in government but also as the attitudes within governments change and as the personnel of government change. To some degree we have therefore attempted to locate the discussion in its historical context and to concentrate not on the transient forms of actual policy measures, but on the longer-lived controversies surrounding them.

Structure of the Cases

The logic that lay behind the contents of the case studies was a view that the policies that countries have in place and their government R&D spending patterns relate to their economic circumstances and concerns, the decision making structures in place in that economy, the prevailing political philosophies, and the attitudes to the role and functioning of the R&D process in the economy. Following this logic the case studies thus explore four main issues:

- the particular economic circumstances of each country and especially the major local economic concerns;
- the structures and institutions involved in the R&D decision making process in each country;
- the overarching political and administrative culture in each country;
- the nature of policies in place and the dominant characteristics of these policies.

As we deal with each of these issues for every country we look at, each case study has a reasonably common structure. An introductory section discusses national perceptions of economic performance with an emphasis upon R&D. It also highlights current political attitudes, aspects of administrative culture and, where relevant, specific peculiarities of the case.

A second section then introduces the main institutions involved in government R&D policy and budgets. Typically the most important are a science and education ministry or ministries, a finance ministry, a ministry for industry, other ministries, the cabinet, parliament, sometimes regional authorities and advisory boards. We discuss the role and responsibilities of these organizations and their interactions. Consultation procedures (formal and informal), advisory structures within which the institutions operate, the existence of pressure groups, lobbying bodies, channels of influence and conflict resolution are also covered. To the extent that it is possible, we attempt to get behind the institutional façade to explore the role of different stakeholders in the decision making process. Our success in this does differ across countries. In some, the institutions are the key; in others, the decision making process is much more hidden and thus much more difficult to track down. A schematic is supplied for each country that illustrates the main actors and their relationships to each other.

In a third section we explore policy philosophies. The purpose here is to see what are the key debates and controversies that surround the R&D decision making process and to consider how these debates relate to political attitudes and views on the objectives of government. At a simple level this may be just that some countries have a belief in the free market whereas others do not. However, we attempt to go deeper and to illustrate how different institutions in the same country may embody different philosophies as regards R&D. For instance, the philosophy governing R&D directed at the advancement of knowledge or social well-being may differ considerably from that governing R&D aimed at strengthening competitiveness or defence power. One of the themes to appear is the differences in attitude between the treasury or finance ministry and the ministry responsible for industry.

As a conceptual toolkit, we find the notions of étatism, corporatism and liberalism particularly useful here. The Anglo-Saxon culture and tradition of policy making is one where liberalism predominates; there is a clear divide between the private and the public sphere. In the French étatist tradition the state is much more expected to give direction to the economy. The German corporatist administrative culture is one of closer interactions between well-organized interest groups in society and the administrative apparatus. In the smaller north European countries like Finland and the Netherlands also corporatist features predominate. These general traits of national administrations appear to feature also in the way R&D policies are designed and conducted.

The next section in the case studies is on R&D policies themselves. Here we discuss general policy orientation (for example, a preference for direct subsidies versus tax incentives), patterns of government R&D support, and other related issues. In this section we also provide some quantitative indicators of the structure of policies in different countries and we address the relationship between national policies and R&D spending and between EU policies and spending. At the end of each case study there is a short concluding section. We leave drawing general conclusions to the last chapter, where we also make an attempt at comparing the main findings.

1.2 SOME BASIC NOTIONS

Before turning to our first case study country in the next chapter, we recall a number of basic notions here and in the following section provide an overview of basic facts to help to provide some context within which to understand the rest of the book.

Science and Technology Policy

While science policy is concerned with the advancement of knowledge without the objective of solving explicitly defined (short-term) practical problems, technology policy is concerned with the practical application of, often scientific, knowledge in a new way with the aim of generating commercial or other benefits. Or, stated more concisely, technology policy is concerned with the application of science to the needs of humanity. Therefore science policy should create the circumstances and facilities that enable advancement of scientific knowledge. This involves the creation and maintenance of public research institutes and universities and the support (financially or morally) of other initiatives that stimulate the advancement of knowledge. Technology policy, on the other hand, tries to influence the decisions of firms to develop, commercialize or adopt new technologies. Technology policy includes the funding of private R&D, the creation of innovation centres and transfer agencies, and the provision of venture capital.

The common argument for government intervention with regard to scientific and technological development is market failure. In the most general terms, market failure occurs when the free market unaided will not produce the welfare optimal outcome. It has for long been recognized that such market failures will be common in R&D activities. Market failure leads to underinvestment in knowledge production. This calls for government intervention in the form of subsidies, regulation or public provision. The most frequently mentioned causes of market failure are externalities, information problems, public goods, risk and uncertainty, and monopoly power:

1. Externalities or wider benefits occur where firms or investors are unable to appropriate all the economic and commercial benefits of the expenditures they incur. Externalities may therefore, from a social welfare point of view, lead to underinvestment. Examples of sources of externalities are information spin-offs from applied research and the

benefits of training to workers. However, externalities can only justify
government intervention where they prevent a project from going ahead
in the most appropriate form from the point of view of the economy as
a whole. There are plenty of examples of projects which exhibit
significant wider benefits but also yield a perfectly satisfactory return
to the investing firm.

2. Lack of appropriate information by market participants distorts the effi-
    ciency of the market mechanism. For example, weaknesses in education
    and training may mean that market participants often do not possess
    enough of the right kinds of information. Also the provision of informa-
    tion on a commercial basis is particularly prone to market failure.
3. Public goods, for instance measurement standards, are products where
    consumption is non-rivalrous. Public goods may or may not be exclud-
    able. In general the private market will not adequately reward the supply
    of public goods and a free market will thus undersupply such goods (and
    the technology to support them).
4. Risk and uncertainty are inherent characteristics of innovation projects.
    Financial institutions and capital markets may lack the information and
    technological knowledge to properly assess the uncertainty attached to
    innovation projects and may consequently be overcautious in providing
    finance. Similarly, capital markets may not always allocate financial
    risks to those best able to bear them. Thus too much of the financial risk
    will be left to the innovating firm, endangering commercial viability. The
    outcome of a project may be uncertain for technological or financial
    reasons. Collaborative research programmes which examine a wider range
    of technological options than is possible for a single firm may provide a
    solution for technological risk. Projects which are nearer to the market
    are usually less prone to technological and more prone to financial (com-
    mercial) risk.
5. Lack of competition and barriers to market entry can inhibit innovation
    by discouraging new firms or by overpricing technology.

Research and Development

The OECD (1993), so called 'Frascati Manual' distinguishes four different
types of research and development activities:

*   *Pure basic research* is theoretical or experimental work undertaken
    primarily to acquire new knowledge of the underlying foundations of
    phenomena and observable facts, without any particular application or
    use in view. It is carried out for the advancement of knowledge without
    envisaging long-term economic or social benefits and with no positive

efforts being made to apply the results to practical problems or to transfer the results to sectors responsible for its application.

- *Oriented basic (or strategic) research* is carried out with the expectation that it will produce a broad base of knowledge likely to form the background of a solution of recognized or expected current or future problems or possibilities.
- *Applied research* is original investigation undertaken in order to acquire new knowledge, directed primarily towards a specific practical aim or objective. Applied research is undertaken either to determine possible uses for the findings of basic research or to determine new methods or ways of achieving some specific and predetermined objectives. It involves considering the available knowledge and its extension in order to solve specific problems.
- *Experimental development* is systematic work, drawing on existing knowledge gained from research and practical experience, which is directed toward producing new materials, products and devices; toward installing new processes, systems and services; or toward improving substantially those already produced or installed.

The Public Knowledge Infrastructure

The public knowledge infrastructure is the whole public framework supporting the generation and diffusion of new scientific and technical knowledge. It is part of the 'National System of Innovation', the elaborate complex of private and public institutions that together contribute to progress in technology. One of the main objectives of science and technology policy is the creation and maintenance of such a public framework. Within this infrastructure three categories of players are important: government policy makers; universities and research institutes; and intermediary organizations.

Policy makers deal with the development of the public knowledge infrastructure, the formulation of the general objectives and the ways to attain these objectives. Generally, science and technology policy is the responsibility of two ministries, usually those of education and economic affairs; sometimes there is only one ministry in charge of science, technology and educational policies. Other ministries are commonly responsible for science and technology policy with regard to their own field of responsibility (for example, the environment, defence, transport, health care).

Universities and research institutes differ in that the former also educate students. Moreover, universities usually do more basic research and research institutes more applied work. The relative importance of universities and research institutes for technological development differs from country to country. In general, the larger the number of independent public research institutes (that is, not linked to a university), the less important the role of the universities with regard to R&D.

Intermediary organizations are all concerned with the transfer of knowledge from the 'knowledge producers' to the community of users. These organizations are vital parts of the infrastructure that are meant to ensure the efficient management and exploitation of new knowledge. Some intermediary organizations are directly linked to universities or research institutes but many countries have independent transfer organizations. Intermediary organizations have tasks like helping companies to look for solutions to technological problems, assisting them with the introduction of new technologies, arousing awareness of technological possibilities and paving the way to knowledge providers.

Types of Science and Technology Policies

Most governments use a whole array of instruments to promote the generation, application, and diffusion of new knowledge. The most important are:

- *financial measures* which grant a specific financial advantage to companies to stimulate innovative behaviour. These could be:
    - (a)   direct subsidies, usually to carry out a specific project. Sometimes the subsidies have to be paid back if the project appears to be successful;
    - (b)   soft loans with an interest rate below the market rate or with other favourable conditions;
    - (c)   tax grants, giving companies a tax advantage if they engage in specific activities, for example the employment of R&D personnel (as is done in Germany, France and the Netherlands);
- *special R&D programmes* aimed at the promotion of technological progress in specific areas that are considered either in the public interest or to the advantage of a country's competitive position. In general, these special programmes consist of a complete set of support measures (financial and infrastructural policies, venture capital, access to research facilities). Although it is often doubted whether governments are good at picking winners, all governments have made their attempts. Generally, their choice of themes is not very original: biotechnology, new materials and information technology are selected by most as technological areas of special importance;
- *special measures aimed at small and medium enterprises (SMEs)* are meant to overcome the specific problems that SMEs encounter in R&D activities (lack of resources and of time, limited awareness of technological possibilities, dependence on one or two suppliers or customers,

inability to build up a research portfolio) and to exploit their possibilities as important creators of employment;

- *provision of venture capital*, which is aimed at high-tech start-ups or fast growers to provide them with the opportunity to attract equity. Constructions differ: the government gives a guarantee to private investors, thereby reducing the risks for those investors, or the government participates in the company itself. Generally this is done through government-owned investment companies;
- *infrastructural policies to improve knowledge transfer* which aim at facilitating the dissemination of new technologies in various ways;
- *international co-operation*, for example, in the EU or through bilateral agreements, which opens up foreign resources of knowledge;
- *regional initiatives* aimed at the promotion of technological development in a specific, mostly relatively disadvantaged, region;
- *development of a legal framework* for the protection of intellectual capital and for setting technical, safety or environmental standards. In this way government helps to reduce the risks of losses for companies that develop or adopt innovations. The legal status of universities and research institutes influences their willingness and possibilities to co-operate with private companies in R&D and to transfer knowledge.

Science and Technology Policy in Relation to Other Policies

Not only science and technology policy influences the decisions of firms with regard to innovation. A number of other policies, although not implemented with the primary objective of affecting innovative performance, may also affect technological development. Among the main policies are the following:

- *Education policy* has as one of its main objectives the creation of a capable workforce. It is hard to overestimate the importance for competitiveness of a well-educated workforce in which there is a balance of academics, engineers, designers and skilled craftsmen.
- *Regional policy* is directed at the development of a specific region, usually one that suffers from a general decline of the dominant industry or from a geographically isolated position. Some regional policies are focused on attracting foreign investment, others on the creation of infrastructure, sectoral development or support of SMEs.
- *Competition policy* aims to ensure free competition in markets. The relationship between market structure and innovation has been a subject of fierce debate among economists. There are two counteracting tendencies: on the one hand, a certain degree of monopoly power

allows firms to mobilize the resources necessary for innovation and helps them to keep the results for themselves; on the other hand, a certain degree of competition forces firms to be innovative in order to keep competitors at bay.

- *Procurement policy* aims to supply the state with the products it needs to fulfil its tasks. It may influence technological development in various ways: (a) it may set clear technological goals for the product or process to be developed; (b) it may set an example; (c) it may enable the innovator to avert some of the development risks and to earn back part of the development costs. Procurement policy is especially important for technological development in the area of military defence. In most OECD countries defence R&D plays a relatively minor role, but in a few countries (the UK, France and the US, but also Israel and Taiwan) defence R&D expenditure approaches or exceeds half of total government R&D expenditure.

1.3   SOME BASIC FACTS

The purpose of this section is to provide some general background information on the countries which will be discussed in the rest of the book. Using OECD data we first consider some general characteristics of the case study countries and then move on to science and technology indicators. The data help to illustrate some of the long-term priorities of science and technology policy at an aggregate level of those countries. Available data cover a period from 1981 to 1995–6 and therefore do not yet reflect the effects of recent developments in R&D policies.

Table 1.1 shows figures on wealth, wealth growth, R&D intensity and (growth of) R&D expenditures. The countries in the list are very different in size. There are small economies (Finland, Israel, Singapore), middle-sized economies (Taiwan, the Netherlands), large economies (France, Germany, the UK) and very large economies (Japan, the US). The populations of France and the UK are almost of equal size. The European countries are roughly similar as far as per capita GDP is concerned. The USA and Japan show a higher figure, whilst Israel and Taiwan score a bit lower. The table shows some significant differences in growth performance over the last 16 years. The Asian countries certainly had the highest growth rates. Israel and Finland went through deep recessions in the 1980s, and the other western countries did not perform very well either during that period. Germany increased in size, but also decreased in per capita income due to the reunification. Taiwan has made very significant progress over the last 16 years but per capita income is still well below that of the other countries. Singapore man-

Table 1.1   Basic statistics

Country	Real GDP[a] (billion) 1996	Population (million) 1996	Real GDP per capita[a] 1996	Growth of GDP (%) 1981–96	GERD[a]/ GDP 1995	Growth of GERD/ GDP 1983–95
Finland	94	5.1	18 400	24.2	2.37	74.3
France	1 267	58.4	21 700	22.8	2.33	10.4
Germany	1 670	81.9	20 400	13.5	2.30	−8.7
Netherlands	325	15.5	21 000	30.3	2.09	5.0
UK	1 174	58.8	20 000	37.3	2.05	−6.4
Israel[b]	93	5.8	16 000	33.8	—	—
Japan	2 991	125.9	23 800	48.5	2.77	17.9
Singapore	78	3.6	21 700	114.0	—	—
Taiwan	283	21.5	13 200	144.1	1.80	—
US	7 576	265.6	28 500	25.2	2.55	−4.1
OECD	19 708	987.5	19 960	—	2.15	0.5

Notes:
[a]   US$, based on purchasing power parity. GDP per capita growth is in constant 1990 prices.
[b]   Israel: 1980–96, civilian R&D expenditures are about 2.33 per cent of GDP (Israel Central Bureau of Statistics).
[c]   GERD = gross expenditure on research and development.

Sources: OECD, *Main Science and Technology Indicators*; *Handbook of International Economic Statistics 1997*. For Israel, Singapore and Taiwan information was used that was made available on the Internet, either by the national bureau of statistics or a ministry (this should therefore be interpreted with care as there might be inconsistencies with the other figures).

aged to catch up with Europe. Apparently there is no correlation between size and economic performance, but size certainly has implications for the way the national innovation system functions. The smaller the country, the more it is forced to set priorities in areas of research and development and to seek international co-operation if a critical mass cannot be attained.

The *World Competitiveness Yearbook 1998* provides information on the competitive performance of the case study countries. Their ranking has changed considerably over the last four years (see Table 1.2). Remarkable is that the smaller countries, with the exception of Israel, perform much better than the larger ones (with the exception of the US).

Table 1.3 shows the relative patterns of specialization in manufacturing of the seven case study countries. It clearly reveals the dominance of France, the

Table 1.2   Ranking of case study countries according to the World
Competitiveness Yearbook 1998

Country	Ranking in 1998	Ranking in 1994
USA	1	1
Singapore	2	2
Netherlands	4	8
Finland	5	19
Germany	14	6
Taiwan	16	22
Japan	18	3
UK	12	14
France	21	13
Israel	25	—

Source: *World Competitiveness Yearbook 1998.*

UK and the US in large-scale industries like aerospace manufacturing, which is also related to their heavy involvement in military production. The dominance of Japan in the electronics and computer industries also clearly shows. Only Finland and the Netherlands export more electronic products than they import. The presence of Philips in the Netherlands and of Nokia in Finland accounts for this. Germany, the UK and France are very strong in pharmaceuticals. Japan, Germany and Finland have a strong position in other manufacturing industries. In the case of Finland most of these exports relate to paper and forestry products.

Gross expenditure on R&D (GERD) is a commonly used indicator of the technological dynamism of a country. It has a number of problems, being a measure of inputs (effort) rather than of outputs (performance) but these are well known and do not need to be repeated here. Figure 1.1 shows the GERD/GDP ratio on the horizontal axis and the percentage growth in this variable on the vertical axis. Indicated are the values for some of the case study countries together with the figures of some other OECD member states. Both axes meet at the OECD averages. Over the period considered the average OECD GERD to GDP ratio hardly changed.

The figure clearly shows some catching-up behaviour. Countries that had a small GERD to GDP ratio in 1983 had growth above the OECD average, and therefore caught up with the leaders (France, Germany, Japan, Sweden, US and UK), some of which had negative growth rates. In 1995 the leaders, with the exception of the UK, still had a GERD to GDP ratio above the OECD average, but they were joined by Finland that now had a rate of 2.37 per cent

Table 1.3 Export–import ratios in 1995

	Aerospace industry	Electronic industry	Office machinery & computer industry	Drug industry	Other manufacturing industries	Total manufacturing
Finland	0.36	1.47	0.7	0.33	1.72	1.58
France	2.35	0.97	0.68	1.25	1.06	1.08
Germany	1.08	0.93	0.59	1.51	1.35	1.28
Netherlands	0.98	1.12	0.9	1.02	1.15	1.11
UK	1.53	0.87	0.97	1.74	0.86	0.9
US	2.75	0.65	0.59	1.12	0.71	0.73
Japan	0.26	3.07	1.96	0.39	1.64	1.75

Source: OECD, *Main Science and Technology Indicators, 1998.*

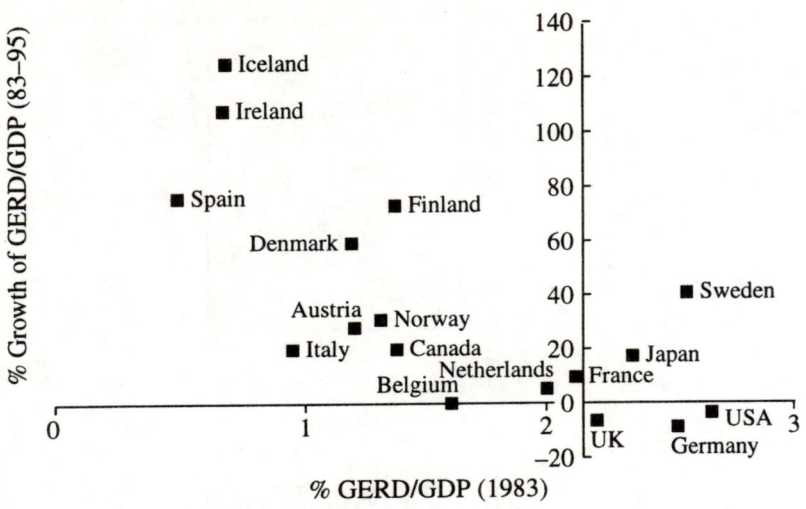

Source: OECD, *Main Science and Technology Indicators, 1998.*

Figure 1.1   GERD/GDP versus growth of GERD/GDP

against an OECD average of 2.14 per cent. So in 1995 the GERD to GDP ratios of the OECD countries did converge. However, the most important developments were taking place outside the OECD, where countries like South Korea (an OECD member since 1996), Taiwan and Singapore were quickly catching up. Rapid developments like these have fuelled worries in countries such as the UK, the Netherlands and Germany about their position in the global race, which we shall discuss in the chapters to follow.

We now turn to some indicators of technological performance, such as the technological balance of payments. The technological balance of payments is reflected by the coverage ratio, a ratio of receipts from other countries for technologies exported and payments made to other countries for technologies imported. The interpretation of technological balance of payments figures is not always straightforward. High imports (a ratio smaller than 1) may mean that a country is poor at developing technology itself, but it may also indicate that it is a particularly keen user of new technology. High exports may mean that a country is a particularly important developer and originator of technology, but it may also indicate that a country is not able to put new technology efficiently to use, such that it is left to others to commercialize it.

For many years the US has been the largest net exporter of technologies (a ratio significantly above 1). However, its technological balance of payments has declined considerably since 1981, indicating that US technological dominance has declined. In 1995 Japan was the second largest net exporter of technologies, while in 1981 it was still a large net importer. The UK is the third largest net exporter, and this is seen as proof of the general concern that the country is good at originating new technologies but poor at exploiting them. Germany has been a net importer of technology for many years now with a ratio that has remained relatively constant at around 0.78. For the other case study countries, recent data are lacking. France was a net importer of technologies in 1992, as it was, to a lesser extent, ten years before. In 1992 the Netherlands still was a small net exporter of technologies.

Patent data are another measure of technological performance. Their disadvantage is that patent counts do not necessarily compare like with like. A common method to overcome the latter problem is to use only patent applications made outside the originating country (external patent applications) on the grounds that this will sift out the minor applications. Figure 1.2 shows external patent applications as a share of total OECD external patent applications. Since 1981 the number of external patent applications has grown by more than 400 per cent (OECD average 1981–95), reflecting the increased globalization of economic activities.

Over the period 1981–95 the USA increased its lead in external patent applications despite, as we saw, a decreased GERD to GDP ratio. In 1995 the

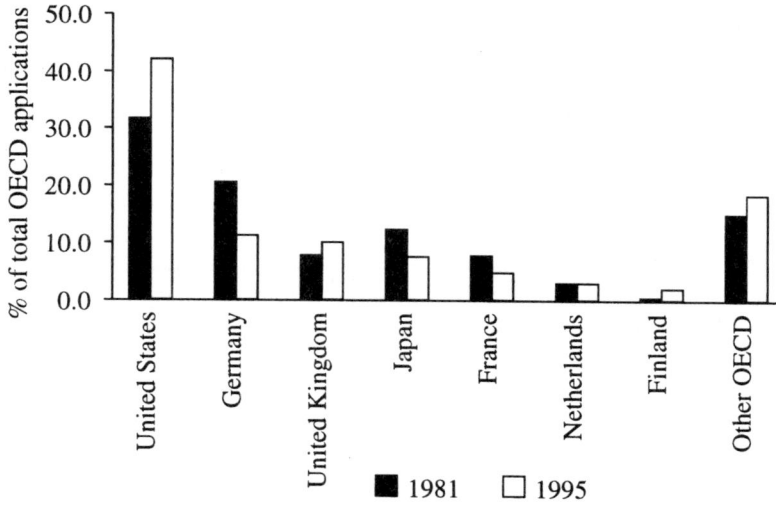

Source: OECD, *Main Science and Technology Indicators, 1998*.

Figure 1.2   External patent applications as a share of total OECD external patent applications

US accounted for 41 per cent of external patent applications. Germany, the next largest applicant, clearly lost ground. The growth of German's external patent applications in absolute terms has been rather modest. The same goes for France and Japan. By contrast, the UK has increased its share of patent applications, although its GERD to GDP ratio is rather low and has been decreasing over a long period of time. Finland, which has increased its GERD to GDP ratio significantly, more than doubled its share of OECD external patent applications. We may conclude, however, that the case study countries, except Finland, are not amongst the best performers in external patent applications. Although the UK and the Netherlands have shown reasonable growth in external patent application, growth in France and Germany has been quite lethargic. Countries like Australia, Sweden, Norway, Canada and Denmark show a much higher growth than Germany and France, but starting from a much lower basis. All countries increased the number of external patent applications per dollar of R&D expenditure, except Japan where this ratio remained fairly constant. Finland quadrupled its number of patents per dollar of R&D expenditure, indicating perhaps an increased awareness among Finnish entrepreneurs and researchers of the importance of the outside world to the Finnish economy. The US and France have a relatively small number of patents per dollar of R&D expenditure, probably reflecting the importance of military R&D in these countries.

One way of judging the degree of government commitment and involvement in R&D is to compare the public share in total R&D expenditures. Table 1.4 gives information on government financed R&D as a percentage of GERD. The share of government funds in total R&D expenditure is on the decline in

Table 1.4   Shares in total R&D expenditures, 1981–95 (%)

	Public		Private		Foreign	
	1981	1995	1981	1995	1981	1995
Finland	43.4	35.1	54.5	59.5	1.0	4.5
France	53.4	42.3	40.9	48.3	5.0	8.0
Germany	40.7	37.4	57.9	60.5	1.0	1.7
Japan	24.9	20.9	67.7	72.3	0.1	0.1
Netherlands	47.2	42.1	46.3	46.0	5.2	9.3
UK	48.1	33.3	42.0	48.0	6.9	14.3
US	49.3	36.1	48.8	59.9	—	—
OECD	45.0	34.5	51.2	59.1	—	—

Source: OECD, *Main Science and Technology Indicators, 1998.*

all countries between the early 1980s and 1995, especially in countries with large defence R&D outlays. In our case study countries the public sector share varies between 33 and 43 per cent, with an OECD average of 35 per cent. Japan is an outlier with a publicly financed share of 21 per cent. Government involvement in R&D is higher in France than in the UK, with Germany in between.

The decrease of government involvement in R&D is of course mirrored by an increase in the share of privately funded R&D. In those countries that are members of the European Union, international funding of R&D is rapidly increasing, especially in the UK and Finland. This illustrates the strong competitive position of UK R&D and the increasing involvement of Finland in EU research programmes. Not surprisingly, international co-operative R&D is a major focus of Finnish science and technology policy.

In a number of countries, military R&D is a major component of R&D expenditures, whereas in others it is negligible. Although the private sector itself will spend money on military R&D as well (but there are hardly any data available on this), the government is the largest financial contributor to it. Defence R&D as a proportion of government R&D spending is highest in the USA, at almost 55 per cent, followed by the UK with 37 per cent and France with 29 per cent (see Table 1.5). The other countries show a much smaller proportion of spending on R&D for military purposes. France and the UK thus differ considerably on this dimension from Germany, where the share is below 10 per cent, and the Netherlands and Finland. The UK and France, on this dimension rather similar to the US, were characterized by Ergas (1987) as mission oriented as compared with countries like Germany which he called diffusion oriented. This strong emphasis on defence R&D,

Table 1.5   Share of defence-related R&D in total government appropriations for R&D (%)

	1985	1990	1996
Finland	1.8	1.5	2.0
France	32.5	40.0	29.0
Germany	11.9	13.5	9.8
Netherlands	3.0	3.0	3.2
UK	48.5	43.7	37.0
Japan	3.8	5.4	5.9
US	67.5	62.6	54.7
OECD	43.4	40.2	35.4

Source: OECD, *Main Science and Technology Indicators, 1998.*

although today less extreme than it was in the 1980s, might have harmed the overall technological performance of these countries, since military R&D competes with civilian R&D for the same scarce resources. There is a tendency in most countries to reduce the share of military R&D in total R&D expenditures, an exception being Japan where military R&D expenditures, though still at a relatively low level, have shown quite strong growth rates.

A considerable part of R&D funded from the public purse is not carried out by public institutions. Table 1.6 shows R&D performed by sector and may be compared to Table 1.4 to see the degree to which there exists a net transfer of R&D funds from the public to the private sector. The 'coverage ratio', the ratio of R&D funded by the private sector to R&D performed by the private sector, has an OECD average of 0.88. Where governments generally transfer R&D funds to the private sector, the Japanese authorities again are an exception with a ratio of 1.03. The Dutch ratio is exactly on the OECD average and Finland and Germany are slightly above it with 0.94 and 0.92 respectively. High public spending on defence R&D, which is often spent on research contracts with private firms, makes itself felt here in the ratios of France with 0.79, the UK with 0.73 and the US with 0.83, which are well below the OECD average.

The importance of public research organizations differs across countries. In some countries they play a major role (for example, in France and the Netherlands), not only in applied research but also in strategic and basic research. In other countries basic and strategic research is the remit of the universities.

Table 1.6   Shares in total R&D carried out per sector (%)

	Private sector		Higher education		Public sector		Private non-profit	
	1981	1995	1981	1995	1981	1995	1981	1995
Finland	54.7	63.2	22.2	19.5	22.5	16.6	0.6	0.6
France	58.9	61.0	16.4	16.7	23.6	21.0	1.1	1.3
Germany	70.2	65.7	15.6	18.9	13.7	15.4	0.5	—
Netherlands	53.3	52.2	23.2	28.8	20.8	18.1	2.8	1.0
UK	63.0	65.5	13.6	18.8	20.6	14.5	2.8	1.2
US	70.3	71.8	14.5	15.2	12.1	9.5	3.1	3.4
Japan	66.0	70.3	17.6	14.5	12.0	10.4	4.5	4.8
Total OECD	65.8	67.3	16.5	17.7	15.0	12.2	2.6	2.7

Source: OECD, *Main Science and Technology Indicators, 1998.*

The OECD average ratio of GERD performed by government is 12.2 per cent. This figure is straddled by the case study countries: Finland, France and the Netherlands have values above, Germany and the UK below the average (see Table 1.7). Apart from Germany, each of these countries appears to show a declining share of R&D being performed by government over the 1981–95 period. The decline is, however, most marked in the UK, where the share has fallen by almost 30 per cent. This reduction reflects a number of policy changes but perhaps more than anything else is the result of the continuing process of privatization of research facilities.

Institutes of higher education have a dual role in society: they teach students and they do research. In both roles they are important to the technological development of a country. Data concerning funds devoted to higher education are not easy to interpret, since it is not always clear to which of these roles funds are dedicated. Nevertheless, differences in government spending on higher education will to some degree reflect differences across countries in their commitment to science, as institutes of higher education are mainly active at the basic and strategic end of the research spectrum. Table 1.7 shows that for most countries expenditures on R&D at institutes of higher education (HERD) are roughly equal to the OECD average of 0.38 per cent of GDP. Finland and Germany have values a bit higher and the Netherlands spends considerably larger amounts. In most countries HERD has increased significantly since 1981, most notably in the Netherlands and Finland.

Table 1.7   Higher education expenditures on R&D (%)

	As percentage of GDP		As percentage of GBOARD[a]	
	1981	1995	1983	1996
Finland	0.27	0.46	29.8	28.1
France	0.32	0.39	16.1	22.5
Germany	0.38	0.43	36.4	41.3
Netherlands	0.43	0.60	45.3	43.8
UK	0.32	0.39	30.7	28.4
US	0.35	0.39	—	—
Japan	0.37	0.40	—	41.4
OECD	0.33	0.38	—	—

Note: [a] General university funds as percentage civil government budget outlays for R&D (Civil GBOARD).

Source: OECD, *Main Science and Technology Indicators, 1998.*

Table 1.8   Government expenditures on R&D by socio-economic objective (%)

| | % of total GBOARD Civil budget R&D | | % of Civil GBOARD | | | | | | | | | |
| | | | Econ. dev. prog. | | Health and environment | | Space | | Non-oriented research | | General university funds | |
	1985	1996	1985	1996	1985	1996	1985	1996	1985	1996	1985	1996
Finland	98.2	98.0	44.8	43.2	18.4	14.8	0	3.0	10.1	10.9	26.7	28.1
France	67.5	71.0	38.7	19.1	12.3	12.5	8.2	15.3	21.0	27.0	17.3	22.5
Germany	88.1	90.2	34.8	23.1	12.0	12.7	4.4	5.5	12.9	16.5	35.7	41.3
Netherlands	97.0	96.8	26.5	25.7	9.5	8.5	2.0	4.3	9.9	12.1	49.2	43.8
UK	51.5	63.0	40.1	16.6	15.6	31.7	5.7	4.3	8.2	18.3	29.9	28.4
US	32.5	45.3	28.7	20.5	42.9	45.1	16.8	25.2	11.5	9.2	0	0
Japan	96.2	94.1	36.1	34.4	5.0	6.9	6.6	7.0	7.7	10.2	44.6	41.4
Total OECD	56.6	64.6	35.6	22.8	20.4	26.3	8.6	13.4	12.5	13.5	0	0

Source: OECD, *Main Science and Technology Indicators, 1998.*

20

Table 1.7 also presents data on university spending as a proportion of civil government budget outlays for R&D (Civil GBOARD). This should provide a rough and ready indicator of government commitment to science as opposed to technology policy. These data indicate a particular emphasis upon university funding in Germany, Japan and the Netherlands, with values over 40 per cent. France, the UK and Finland all show values below 30 per cent: the UK and Finnish ratios having fallen since 1982. In the Netherlands there is some concern whether its spend on universities is not devoted too much to non-technical disciplines as the high level of the spending rate is not reflected in Dutch technological performance.

There are significant differences across countries in the emphasis placed upon different socio-economic objectives when comparing the more recent data. The first and most obvious difference is with respect to defence. We noted this above and indicated that on this dimension France, the UK and the US were very different from Germany and Japan (and the Netherlands and Finland). Looking at the share of different programmes in total civil R&D expenditures of the government, we also notice some differences. Economic development programmes have remained relatively important in Finland and Japan over the years. These programmes became significantly less important in France, Germany, the UK and the US. Economic development programmes cover R&D for the purpose of the advancement of agriculture, fishery, forestry, industry, energy, telecommunications, transport and urban and rural planning. By contrast non-orientated research became more important in most countries.

APPENDIX: *PRO FORMA* QUESTIONNAIRE

The questions below refer to two main topics: objectives of science and technology policy and mechanisms of policy generation. Our emphasis will be on the latter. The questions do not address characteristics of budgets in particular as most of this information is available through published documents.

The interview focuses on the objectives and mechanisms of science and technology policy as a whole and of government financing of R&D activities in particular. The attempt is to put government R&D budgets into perspective by tracing an overarching 'policy vision' and by identifying links to other policies. We hope to obtain an impression of these issues, especially of the policy generation mechanisms, as they have developed in their historical context.

Objectives and Instruments of Science and Technology Policy

1. *What is the 'mission' of your organization with respect to science and technology policy?*

    (a)   What are objectives and priorities in science and technology policy (what is the weight of: economic competitiveness on international markets, social and ecological objectives, national security)? How are these objectives operationalized? Are these objectives seen as compatible or conflicting?

    (b)   Is the main focus of policy on: innovation or diffusion (exploitation); general advancement of technological capabilities or targeted advances (for example, in areas of particular strength)?

    (c)   To what extent is it a task of your organization to contribute to: creating conditions; identifying direction; taking specific initiatives?

2. *Is your organization's policy a response to:*

    (a)   'market failure' (dynamic inefficiency): in private sector innovative activity (generation); in market driven diffusion of innovations (exploitation)?

    (b)   'institutional failure', that is, weaknesses in: systems of education and training (skills shortages); systems of financing business activity (short-termism); methods of management (strategy) and organization of businesses?

    (c)   'structural' weaknesses in the national economy: economic structure (weaknesses in specific 'strategic' industries/dominance of mature industries; overrepresentation of defence-related industries); technological base (revealed technological advantage, labour and total factor productivity, labour skills and levels of education); environmental opportunities and pressures?

    (d)   'foreign policies': matching of foreign initiatives, 'tit for tat' or 'level playing field' arguments?

    (e)   'technological opportunity': to generate technologies that will find a market in the future (technology push)?

3. *What are the main instruments to attain these objectives?*

    (a)   Regulation: for example, domestic content requirements; financial incentives: subsidies, taxes, risk sharing mechanisms; public procurement; development of an information infrastructure; promotion of collaborative networks?

(b)   Is there a preference for subsidy over tax instruments; if so, for what reasons?

4.   *How is science and technology policy related to other policies?*

Competition policy; education policy; trade policy; policy with respect to (inward) foreign direct investment.

5.   *To what extent are objectives of national science and technology policy co-ordinated within the European Union?*

How does EU technology policy impact upon national technology policy?

Mechanisms of Government R&D Budget Setting

1.   *What is the organizational structure within which government R&D budgets are determined; who is involved?*

(a)   What is the position of your organization; what is its function?
(b)   What government departments are involved; what is the relation of your organization to these departments?
(c)   What other organizations are involved, either as representatives of stakeholders or in an advisory function?
(d)   How are the following interests represented in the decision making process: large domestic/foreign businesses (lobbies); public enterprises; SMEs; labour; geographic regions; universities and science; government research institutes; international organizations; capital markets?
(e)   Which levels of government are involved (central, regional, local) and in what way?

2.   *How does this organization operate; what are its procedures?*

(a)   What are the stages at which technology policy is generated?
(b)   Who takes initiatives (in the public sector; in the private sector)?
(c)   What channels of communication are important (for example, between departments or between them and outside organizations)?
(d)   How are responsibilities and competencies distributed (for example, between different departments)?

3.   *How is the national government represented in the decision making process at the European level?*

4. *What is the relationship between domestic spending on science and technology and the national contribution to European spending?*

    (a) How and by whom is this contribution to European spending decided?
    (b) Is a principle of non-additionality operative; how does it function?

5. *How does allocation of funds to potential recipients take place?*

    (a) Are there general guidelines with respect to funds allocation mechanisms (tenders, etc.)?
    (b) What is specified in the typical contract between government and recipient (deliverables, risk of failure, phasing, ownership and use of results)?
    (c) How are supported projects monitored; does monitoring play a part in project management?
    (d) How is your organization's policy evaluated?

Evaluation/Future Trends

1. *Does current government policy with respect to financing R&D attain its objectives?*

2. *Do the present mechanisms in place to determine this policy function adequately?*

3. *How is government policy regarding spending on R&D likely to evolve in the future?*

    (a) How is science and technology policy affected by the process of European integration?
    (b) How is science and technology policy affected by the general trend toward privatization?
    (c) What other structural changes may impact on the future development of science and technology policy?

Perception of the European Context

1. *What do you perceive as the main differences between the approach taken in your country and the approaches taken elsewhere in Europe?*

    What are important differences in conditions; what are important differences in objectives; what are important differences in mechanisms?

REFERENCES

Ergas, H., 1987, 'The Importance of Technology Policy', in P. Dasgupta and P. Stoneman (eds), *Economic Policy and Technological Performance*, Cambridge University Press, pp. 51–96.

International Institute for Management Development, 1998, *The World Competitiveness Yearbook*, Lausanne, Switzerland. Also available on the internet: www.imd.ch

OECD, 1993, *Proposed Standard Practice for Surveys of Research and Experimental Development (Frascati Manual)*, Paris: OECD.

OECD, 1997, *Main Science and Technology Indicators*, Paris: OECD.

2. The United Kingdom

2.1 INTRODUCTION

From being the first industrializing economy and the world leader in trade and manufacture the United Kingdom (UK) is now only one of many advanced players in the world economy. In terms of productivity, technological performance, economic prosperity and other similar indicators the UK rarely ranks in the first five (or often the first ten) in international league tables. The economy has undergone much reorientation since membership of the European Community was agreed. Trading patterns have shifted away from the old Commonwealth and the US towards new European markets (although the growing importance of Japan as a trading nation in the postwar period means that Japanese imports are very significant in certain fields, especially electronics). In the early 1980s, with the prospect of North Sea oil and the newly installed Conservative government, the resulting movements in exchange rates had a decimating effect on the UK manufacturing sector, with large-scale redundancies and firm closures. Not until the early 1990s did manufacturing output return to its 1979 levels. The reduction in size by about one-third was argued to have left a 'leaner and fitter' manufacturing economy. In international league tables UK productivity growth in manufacturing is now nearer to the top. However, there is considerable scepticism in UK academic circles as to whether this is any more than a statistical mirage. With the declining size of the manufacturing sector and a very small agricultural sector, the service sector is of prime importance in the UK economy, providing almost 70 per cent of employment.

The UK is a very open economy and therefore it is particularly concerned with international competitiveness. Too often, however, this competitiveness relies upon low labour costs rather than high productivity or good product design. By comparison with its international competitors the UK is a low-wage economy. In 1995 UK hourly labour costs in industry were less than US$14, compared with 17 in the US, around $24 in the Netherlands and Japan, and almost $32 in Germany (EZ, 1997). All opinion makers would prefer to see the economy as one in which high productivity could support high wages.

The UK has both promoted itself as a home for and been willing to react favourably to foreign multinationals wishing to establish manufacturing, service and R&D facilities in the economy. It has been particularly successful with Japanese and US companies in the motor vehicles and electronics sector. In the chemicals sector many international companies undertake R&D in the UK. This has been given some encouragement by such companies' fears of the 'Fortress Europe' concept. However, UK-based multinationals are also major investors abroad with both production and R&D facilities around the globe (some would argue at the expense of the home economy). The role of the City of London as an international financial centre also acts as a factor influencing UK attitudes towards capital mobility.

The UK is a clear example of a very concentrated economy with a small number of large firms dominant in both production and R&D. In R&D the top 20 companies account for around 55 per cent of the total private R&D spend, and 500 out of a total of around 5700 do practically all of private R&D (DTI, 1997b). The largest spenders are firms in the pharmaceutical and chemical industries. There are a number of large players with a defence orientation in aerospace, electrical machinery and telecommunications equipment (UK R&D Scoreboard 1997). SMEs form only a small part of the total UK production base (in manufacturing) especially compared to, for example, Germany, although after the early 1980s recession and in the face of continued government support the SME sector has grown in importance over the last decade.

The dominant political culture in the UK is primarily determined by the party in power. There is no tradition of coalition governments: the governments of the last 50 years have been either Labour (that is, socialist) or Conservative. The party in power nearly always has a significant parliamentary majority. The administrative (civil service) culture is dominated by the political culture and civil servants primarily act as directed in general or in particular by Ministers. Civil servants on policy making grades are largely generalists rather than specialists. There is no tradition of consensus seeking or consensus politics in the UK (although this does not rule out consultations). With a single party in government and an official opposition the regime may be best considered as confrontational.

The UK had a Conservative government in power from 1979 until 1997 (with three election victories and one change of leader). In 1997 a Labour government was installed. The preceding Conservative regime had been particularly radical with strong (but weakening) right-wing attitudes. The dominant attitudes of the government over this period were:

- a belief in the need to roll back the frontiers of the state through tax reduction, privatization, reduced government intervention and a tightening of social security coverage (thereby making citizens carry greater responsibility for their own actions). It was further believed that government (that is, civil servants) are not well equipped to make commercial judgements;

- an overriding belief in the efficacy of the free market and thus a belief in the advantages of competition. This has been reflected in, for example, deregulation (as well as privatization), a withdrawal from support of near-market R&D and a desire to improve the market environment. It has also been the argument put forward to justify the considerable reduction in the power of trade unions in the UK economy. (Conservative governments have always been caricatured as pro-business/anti-union; Labour governments as pro-union/anti-business.) Union policies in the UK used to be radical in comparison with those in such countries as Germany and the Netherlands, thereby fitting in with the overall confrontational political culture. Compared to the situation in the 1970s, unions are now very weak and play very little role in any government policy making processes, although under the new Labour government the voice of organized labour is again being heard;
- a belief that macroeconomic policy should be dominated by the need to maintain low and stable rates of inflation, which in turn can be best achieved if the Public Sector Borrowing Requirement is strictly controlled. It was also believed that hitting the inflation target would in due course necessarily lead to improvements in the supply side of the economy, especially in R&D and competitiveness;
- an ambivalent attitude towards further European collaboration and cooperation. In addition to concerns over national sovereignty this reflects in no small part a belief in the UK that recent hard-won changes, based upon the above three beliefs, would be threatened by Community-wide policies based upon the more liberal attitudes of European partners (which is no better reflected than in UK attitudes to the Social Chapter).

Free-market beliefs and attitudes were pervasive within government when the Conservatives were in power and have now become established as dominant in the economy as a whole. So far, as elsewhere in Europe, Labour taking over has not led to any widespread overturning of these general beliefs and attitudes. Market principles are the overriding factors not only in industry but in the science (university) sector as a well and, moreover, are also dominant in the functioning and evaluation of government agencies and departments.

The two main political parties have very similar objectives with respect to the need for improvements in UK economic performance and the role to be played by R&D. The differences (although these have perhaps narrowed in recent years) reflect more the means by which the objectives can be best attained than the objectives themselves. Major policy changes are largely driven by political attitudes rather than reasoned argument.

2.2   STRUCTURES

Private business funds about 47 per cent of total domestic R&D (see Table 2.1), but it performs about 75 per cent of total domestic R&D. In 1985 the UK government still funded 44 per cent of all R&D in the UK, but by 1996 this had gone down to only 33 per cent. While gross expenditure on R&D in the UK went down from 2.23 per cent of GDP in 1985 to 2.05 per cent in 1995, government expenditure on R&D as a percentage of GDP fell from 1.08 per cent (on average between 1981 and 1985) to 0.68 per cent in 1995 (OECD, Main Science and Technology Indicators). The fall in government expenditures on R&D resulted from an increase on spending on the science and engineering base by 14 per cent, a decrease in defence spending of 29 per cent and a downturn in other spending by the civil departments of 39 per cent.

Table 2.1   Overview of domestic R&D spending in the UK in 1996

	Expenditures (£ million)	% share
Government departments	2 445	17.1
Research councils	1 092	7.6
Higher education funding councils	1 027	7.2
Higher education institutions	120	0.8
Business enterprises	6 786	47.3
Private non-profit	546	3.8
Abroad	2 323	16.2
Total	14 229	100.0

Source: Office of Science and Technology, *Science, Engineering and Technology Statistics 1998.*

Looking at where government-funded research is performed: in 1986–7, 28 per cent was performed within government departments, 23 per cent in universities and 40 per cent in private industries and public corporations. In 1996–7 (now including the National Health Service), the percentages are 35 per cent in government itself, 31 per cent in universities and 28 per cent in private industries and public corporations. The shift away from private industry is even more pronounced than these figures suggest because over the last couple of years a number of government research institutes have been given 'agency status' and are now operating under a market regime.

   The main players in the public R&D process in the UK are the Department of Trade and Industry (DTI), the Ministry of Defence (MOD) and the Treasury. The DTI includes the Office of Science and Technology (OST), which takes care of the science budget. The budgets of the Higher Education Funding Councils are the responsibility of the Department for Education and Employment (see Table 2.2).

Table 2.2   Net government expenditures on science, engineering and technology by departments in 1997–8 (estimated outturn)

	Expenditures (£ million)	% share
Science budget, including:	1338	22.5
Engineering and Physical Sciences Research Council (EPSRC)	386	6.5
Medical Research Council (MRC)	289	4.9
Particle Physics and Astronomy Research Council (PPARC)	198	3.3
Biotechnology and Biological Research Council (BBSRC)	188	3.2
Natural Environment Research Council (NERC)	168	2.8
Other	109	1.8
Higher Education Funding Councils	1070	18.0
Total civil departments, including	1361	22.9
National Health Service	400	6.7
Department of Trade and Industry (excl. OST and Launch Aid)	350	5.2
Department of the Environment, Transport and the Regions (DETR)	154	2.6
Ministry of Agriculture, Fisheries and Food (MAFF)	146	2.5
Department for Education and Employment (DfEE)	86	1.4
Others	225	4.5
Ministry of Defence (MOD)	2178	36.6
Total science, engineering and technology	5946	100.0

Source: Office of Science and Technology, *Science, Engineering and Technology Statistics 1998.*

Office of Science and Technology

The Office of Science and Technology (OST) was established in April 1992 to provide a central focus for the development of government policy on science and technology. Until July 1995 it was part of the Office of Public

Service and Science (OPSS) within the Cabinet Office and was overseen by its head, the Chancellor of the Duchy of Lancaster. Since then the OST has been moved to the Department of Trade and Industry (DTI), although within the DTI its functions are supposed to be ring fenced. Ministerial responsibility for all OST functions moved to President of the Board of Trade (that is, the Minister of Economic Affairs).

The OST is headed by the Government Chief Scientific Adviser, who is responsible for giving advice on science, engineering and technology (SET) matters to the Prime Minister, to the minister directly responsible for SET policy (the President of the Board of Trade, as the Minister for Trade and Industry was until recently known), and to other government ministers on transdepartmental SET issues. The Cabinet as the main executive authority plays a co-ordinating role with regard to science and technology policy via the Cabinet Committee on Science and Technology. This Committee was chaired by the Prime Minister until July 1995, and then by the Deputy Prime Minister. The OST takes an overview of the science and technology components of departmental spending and as its head the Chief Scientific Adviser provides advice to this Cabinet Committee on how the numbers add up.

There are two groups within the OST (see Figure 2.1). The first, the Transdepartmental Group, has the responsibility for co-ordinating SET policy across departments (SET is defined to include R&D plus technology transfer and scientific and technical postgraduate education). Thus, the OST oversees the totality of SET spending, but does not have executive authority. Its main tasks are to ensure that there are no major holes in the science and engineering base (SEB) portfolio and that no wasteful duplication of R&D takes place. Its main activities are co-ordinating, taking care of communication between departments and guarding cost-effectiveness. The second group, the Science and Engineering Base Group, has responsibility for the science budget, which mainly provides funds for the Research Councils (but also for the Royal Society and the Royal Academy of Engineering). The six Research Councils form the main channels through which the government supports the UK science and engineering base. The Research Councils spend most of their budgets on research in universities. The President of the Board of Trade appoints their heads. The main objective of the Research Councils is to maintain the dynamic capability of the SEB to generate new knowledge, but also to grasp new ideas, whether generated in the UK or overseas. The UK carries out 5.5 per cent of the world's research effort, producing 8.0 per cent of the world's scientific publications and has a share of 9.1 per cent in the world's citations (DTI, 1997c). It develops maybe 5–10 per cent of technology in the world. The ability to assimilate and to build upon this is important, as is the ability to develop niche specialities. An important argument in constructing an R&D portfolio is the establishment of a reasonable balance of activities.

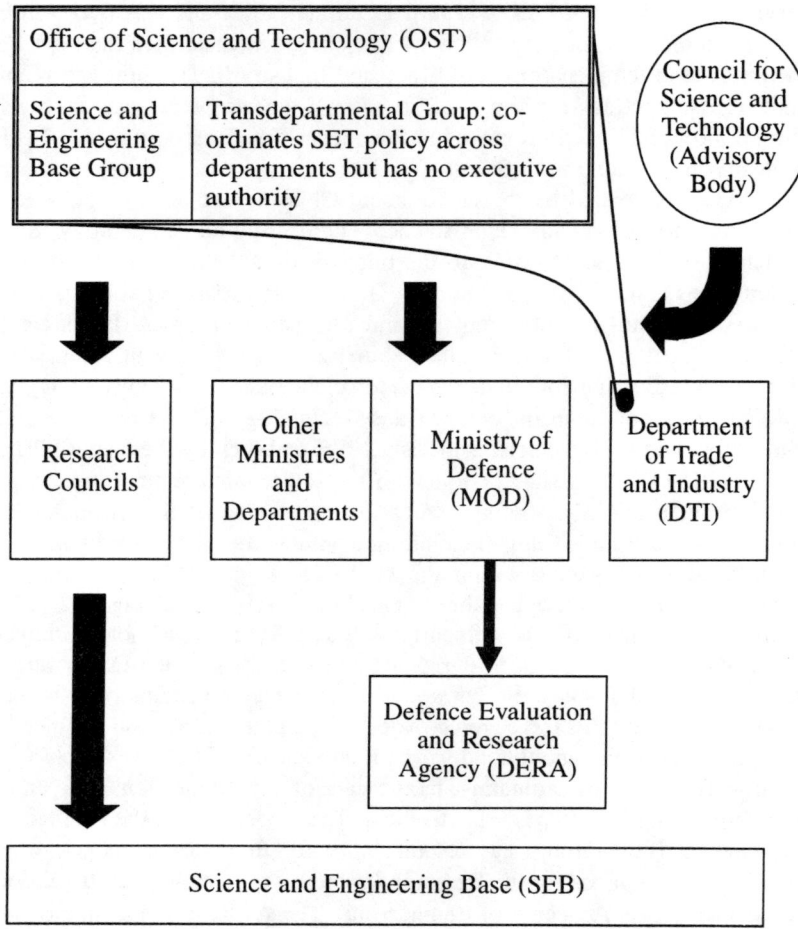

Figure 2.1   The structure of the UK public system of science and technology

In general, projects in the science budget have a long time horizon; our informants argued that it is important not to neglect totally areas that may deliver returns only in the very long term, such as astronomy (although such activities are very technology based and may thus indirectly generate useful spill-overs in the shorter term). There were suggestions that the move of the OST to the DTI will induce a gradual shift of priorities in the allocation of the science budget away from longer-term projects in the direction of shorter-term ones. The scientific community saw the move as an indication that scientific development and economic objectives will be linked even more closely in the future than would have been the case if the OST had remained in the Cabinet Office. There are fears that, by going even further down this road, the unpredictable advantages that come from 'blue sky' research will be driven out of the system. There are no indications so far from the DTI that this is happening. Nevertheless, it is tempting to interpret this change of organizational structure as part of the drive to tie in SET policy as closely as possible to the government's objective of fostering industrial competitiveness and creation of wealth through R&D policy.

Other tasks of the OST are the carrying out of the Technology Foresight programme (see box) and administering the secretariat of the LINK action scheme, an instrument to support collaboration between the SEB and industry. The major players in LINK are the DTI and the Research Councils. The OST furthermore co-ordinates policy across education departments (that is, the Department for Education and Employment, the Northern Ireland Education Department and the Education Departments within the Scottish Office and the Welsh Office) and the HEFCs, and it is responsible for the Public Understanding of Science Programmes. All negotiations about the OST budget during the public expenditure round are conducted at ministerial level. The Director General of Research Councils advises the minister responsible, now the President of the Board of Trade, on the science budget. The President of the Board of Trade then negotiates the science budget with the Treasury.

TECHNOLOGY FORESIGHT

An important new instrument in the determination of new policy directions is the Technology Foresight programme. The primary objective of this programme is 'to identify national and global market opportunities for UK industry over the next 10–20 years, together with the underlying science, engineering and technology investments that will help industry exploit those opportunities, and to foster the creation of new networks between industry, the science and engineering research communities and Government'. The programme was co-ordinated by the Technology Foresight Steering Group and managed by the Office of Science and Technology (OST); within the programme 15 sector panels were set up.

The sector panels and the Steering Group published recommendations in the spring of 1995. Some 27 generic or cross-sectoral areas of SET and some 18 priorities relating to support for infrastructural issues were identified:

A   The Skills Base
1. training better SET teachers;
2. improving communication skills of managers, scientists and engineers;
3. increasing IT competence at all levels;
4. increasing public understanding of science;
5. improving business awareness through education and training.

B   Research in the Science and Engineering Base
6. maintaining support for excellent basic research;
7. new incentives for multidisciplinary research;
8. incentives for universities and research councils to work with industry;

C   Communications
9. promoting IT superhighways;
10. monitoring and disseminating SET intelligence from overseas.

D   Finance
11. development and encouragement of long-term finance for R&D and innovation, including continuous review of fiscal measures;
12. incentives for SMEs: networks, technological incubators;
13. supporting infrastructures which promote successful business clusters (e.g. Silicon Glen).

E   Policy and Regulation
14. intellectual property rights;
15. public procurement as a stimulus to leading-edge technology;
16. regulations regarding environment, finance, communications;
17. standards concerning environment, genetic manipulation, health and safety;
18. demonstrator projects.

The Technology Foresight programme tried to identify (a) where technology is going; (b) where the commercial opportunities are; and (c) the abilities to exploit these. Identification of technological opportunity is an important trigger to action, but only in combination with identified user benefits or market potential. The move towards linking science to wealth creation through the determination of national priorities can, in principal, lead to a technology strategy for all government R&D spending. Some movement in this direction has begun to be perceivable. On the one hand, it does not seem to be the case that the DTI is currently specifically targeting areas in the technology spend that arise from the Foresight exercise. On the other hand, there have been attempts to use the Technology Foresight exercise as a basis for influencing the private R&D spend, thereby creating an economy-wide science and technology strategy. The OST reports that in 1997 about 70 per cent of the research activities in the Research Council portfolio were in Foresight priority areas.

The government intends to continue the Foresight programme and will therefore retain the organizational structure created (sector panels, Steering Group, and so on) and hopes that this will be a basis for collaborative networks within industry and the science and engineering base. The government plans to develop, with industry and the science and engineering base (SEB), collaborative programmes in areas of promise identified by Foresight. Consultation regarding a second major exercise has taken place in 1998.

Ministry of Defence

The Ministry of Defence (MOD) spends about 10 per cent of its total budget on R&D, and accounts for about 37 per cent of R&D expenditure by the UK public sector, down from about 50 per cent in the early 1980s. Of its estimated budget of £2.121 billion in 1997-8, about 28 per cent is spent on applied research (up from 18 per cent in the early 1990s) and the rest on experimental development. Of gross expenditure, about 31 per cent is spent intramurally; most of the balance goes to private industry. Less than 1 per cent (down from 7 per cent in 1989-90) is spent overseas but more than 20 per cent is expenditure on international collaborative projects. The MOD spends about 10 per cent of its total budget on R&D. Its £2.1 billion SET expenditure has underpinned a yearly £5 billion of equipment procurement as well as around £5 billion of defence export sales.

The MOD undertakes significant development in support of equipment procurement. Equipment development is usually undertaken by industry, sometimes as a private venture, but much more commonly funded by government development contracts or combined development and production contracts. Government finances 53 per cent of business defence-related R&D (with half the balance coming from overseas and half from business's own resources), whereas it only finances 5 per cent of civil business R&D. Total intramural defence-related R&D expenditure by business enterprises has declined by more than a third in real terms since 1990. Whether MOD pursues a developmental or an off-the-shelf procurement strategy is decided case by case. However, there is an increasing trend toward off-the-shelf procurement. Over the last few years there has been a changing relationship between the government and its defence suppliers in that contract terms are shifting greater proportions of the risk to the suppliers. There is also a trend towards growing involvement in international defence R&D collaboration, induced by increasing costs and complexity of weapon systems.

The most important recent structural change in the MOD's SET activities has been the setting up of the Defence Evaluation and Research Agency (DERA). In DERA most of the department's non-nuclear SET establishments have been brought together. DERA is managed as a trading fund, introducing formal customer–supplier relationships into the production of research. This mechanism is imposed, on the one hand, to drive down costs and improve efficiency and, on the other hand, to increase the agency's responsiveness to departmental needs and opportunities that might arise in the civil sector.

One process whereby defence technology is transferred to the civil sector is through contracting out research by DERA (or its smaller predecessor DRA). Another is through the establishment by DERA of Dual Use Technology Centres (DUTCs). DUTCs are set up to provide a two-way technology flow, that is, to facilitate the exploitation of DERA's defence technology for civil purposes as well as to provide DERA with access to a wider industry base. The centres currently operating deal with structural materials, supercomputing, marine technology, telecommunications and information processing, software engineering and electronics.

Department of Trade and Industry

The DTI, which since 1992 includes most of the former Department of Energy, spends about 6 per cent of its budget on R&D. However, in real terms its SET budget has decreased by more than 60 per cent since the middle of the 1980s. The DTI administers or participates in numerous R&D support schemes and initiatives that can be grouped into the following six broad categories: industrial innovation; aeronautics; space; support for statutory, regulatory and policy responsibilities; nuclear energy, and non-nuclear energy. In addition the DTI promotes a number of EU initiatives in the UK. These support schemes consume roughly four-fifths of the DTI's innovation, technology and standards (ITS) budget. Currently there is a trend to direct this budget increasingly toward the exploitation of technology and technology transfer. The DTI sets out to develop: a more supportive climate for innovation; greater awareness of best practice; more accessible external advice and services; improved acquisition and use of technology produced elsewhere; and closer relations between academia and business.

FACILITATING TECHNOLOGY TRANSFER

An important government-wide mechanism to support industrial innovation to which DTI is a major contributor is LINK. It supports precompetitive research partnerships by bringing together industrial and science and engineering (SEB) participants. By 1997 over 570 LINK projects, worth in excess of £300 million, had been initiated involving over 800 companies and 130 SEB institutions. Government contributions to LINK amounted to £43 million in 1997–8.

LINK is managed by means of research programmes, which cover discrete technology or generic product areas. Government provides up to 50 per cent of the total eligible costs and industry provides the balance.

The main new vehicle to encourage technology transfer in England is the network of Business Links, local centres providing support services to business, especially small and medium enterprises (SMEs). Similar centres have also been set up in Scotland, Wales and Northern Ireland. The primary function of Business Links is to provide advice on business best practice (regarding management, education and training, finance, technological innovation, and so on) and to act as a delivery mechanism and contact point for DTI support for business generally. Advice on technological matters is via Innovation and Technology Counsellors (ITCs) supported by NEARNET and SUPERNET, local and national networks of expertise around Business Links that are actively being developed. ITCs will put firms in touch with DTI Technology Transfer schemes, but the extent to which such schemes will be delivered through the ITC will vary with the nature of the scheme in question. By March 1995, 75 Business Links were open (up from three at the beginning of 1994). Over 30 of those now have appointed Innovation and Technology Counsellors, experienced people with a technical and business background. Design Counsellors are being appointed and Innovation Credits are being piloted.

The Consultancy Brokerage Service administers a database of detailed information on consultants and their track records and provides firms with free help in selecting consultants to match their needs. Awareness of best practice is promoted through the Managing in the 90s programme, which has already reached over 100 000 people through seminars and workshops, and through the Environmental Best Practice Programme.

The Teaching Company Scheme (TCS), the Postgraduate Teaching Partnership (PTP) and the Support for Industrial Units Scheme (SIUS) aim at technology transfer from the science base toward businesses. TCS involves young, highly qualified graduates working in companies for two years on technology transfer projects jointly supervised by academics and industrialists. PTPs involve groups of 'associates' carrying out practical research projects whilst receiving industrially relevant training within a Research and Technology Organization (RTO). The SIUS aims to strengthen the commercial activities of industrial liaison units in higher education institutions.

Finally, technology transfer from foreign sources is promoted through the department's Overseas Technology Services. Through the Overseas Science and Technology Expert Missions (OSTEMS), support is given to small groups of senior specialists in industry to pay short visits abroad to study a specific technological development. The Overseas Technical Information Service (OTIS, run by Pera International) makes available documents provided by SET posts at UK overseas embassies. The Engineers to Japan Scheme sends engineers in senior management positions to work in Japanese companies.

DTI has been actively involved in the Technology Foresight programme since its inception, working closely with the OST, in order to encourage industry to participate and to ensure that industry's interests are fully taken into account. The DTI was represented on the Technology Foresight programme steering group and on 12 of the 15 sectoral panels.

DTI used to incorporate four government laboratories, but these have been privatized. Their change in status was designed to make them more efficient and to better serve the needs of industry, the DTI and other customers. To the extent that this distances them from the DTI there is a risk that in time they will be less accessible as an informal source of technological expertise.

Other Players

Individual departments or ministries have considerable autonomy in the determination of the level of spending upon R&D and the policies that they effect to stimulate technological innovation. They are, however, constrained by a total departmental spending figure (across all activities, not only R&D) that has to be agreed each year with the Treasury and confirmed by the Cabinet. This 'frame budgeting' approach is found in other case study countries as well.

The Treasury (and its Minister, the Chancellor of the Exchequer), in addition to being involved in the determination of the departmental total spends, has fiscal control and thus control of the tax treatment of R&D. Until very recently there were no plans for special tax incentives for R&D in the UK. Tax incentives were not favoured by the Treasury for a number of reasons. First, the taxation system should be as simple and straightforward as possible and should as little as possible distort the operation of markets. Second, tax incentives are indiscriminate. Distribution of public expenditure, for example, through Research Councils, enables much better targeting. The cost-effectiveness of tax instruments is a problem, as it is difficult to avoid creating loopholes and to monitor accountants. However, there has always been an ongoing discussion with the Treasury, in which OST is involved, about using tax incentives as a way to offset short-termism in financial markets. With the incoming of the current Labour government, the issue of tax incentives has been put once again on the agenda.

THE PUBLIC EXPENDITURE SURVEY

Around the beginning of July the Cabinet meets to determine a public expenditure strategy and decides an overall total of public expenditure to aim for over the next three years. Then, during the summer and autumn, ministers collectively decide how these agreed totals should be divided up between spending departments. There are 'bilaterals', discussions between the Chief Secretary to the Treasury and individual ministers, producing a two-page paper which goes to a Cabinet Committee. This Cabinet Committee is charged with advising the Cabinet on how the approved total should be divided up. It meets the majority of

spending ministers and they put their cases to the Cabinet Committee. The Chief Secretary to the Treasury gives the Committee
his view about the relative priority of a department's expenditure.
The Cabinet Committee then makes recommendations to the
Cabinet. Decisions are taken in October or early November,
which are reflected in a unified budget delivered by the Chancellor of the Exchequer in late November, when the government's
decisions on both expenditure and taxation are announced.

Previously, there was first a spending round and then a decision
process on how to finance expenditure; now, financing plans are
worked out by the Treasury in parallel with spending plans in a
much more integrated process. The key figure is the 'bottom line
total' which functions as a starting point for discussion. Thus, at
the end of November a combined statement is made on departmental spending and taxation.

The Treasury is divided into sections that correspond to spending departments. These sections negotiate with departments about
their budget on an official level. Individual departments and the
Office of Science and Technology (OST) determine on what they
spend research budgets, but the Treasury is interested in the
broad objectives that are being set, to the extent that in the
public expenditure round the Treasury may want to argue for
making savings if the Treasury perceives that money is being
spent on research that does not meet the broad objectives which
the department is pursuing. Civil servants in the Treasury advise
the Chief Secretary on the scope for any economy within the
R&D budget consistent with the achievement of government objectives and on the extent to which the R&D budget is contributing
to wider national economic objectives.

The efficiency of the public expenditure survey process is not
formally evaluated but there is a debrief every year after the
survey is completed at which ministers collectively consider
whether steps for improvement of the process can be taken. In
the Treasury attempts are made to develop a more strategic
approach, to have a more constructive relationship with departments. That does not mean giving up aiming for economy, but
implies striving to understand departments' objectives and develop a strategic view from a Treasury vantage point regarding
the contribution R&D expenditure makes to national wealth. On

the whole, from the perspective of the Treasury, the mechanisms seem to function well. For example, there is a healthy balance between, on the one hand, OST, which has the purpose of considering and marshalling the arguments for public investment in R&D, and, on the other hand, the Treasury, which acts as a necessary constraint and asks hard questions about the claimed returns.

Responsibilities for manpower, training and skills in the economy were until recently the responsibility of the Department of Employment. Paradoxically, however, given the continuing commentary upon the deficiencies of the UK economy in terms of skills, the DoE was not usually considered in the UK as part of the R&D process. Since late 1995 the Department of Employment has been merged with the Department for Education to form the Department for Education and Employment (DfEE). The Department for Education had responsibility for primary and secondary education and is primarily responsible for the higher education budget (which forms part of the university budget). The merger thus brings manpower and education closer together. There also now seems to be much more discussion than in the past between the DTI (and the OST) and the DfEE regarding shared issues.

The role of Parliament in the determination of R&D budgets and policies is apparently an indirect one. Parliament needs to approve the total government spending and tax plans each year via the Budget process, but this is largely a matter of agreeing the totality of government spending and tax revenues rather than a consideration of the detail of government spending plans of individual ministries. Through the House of Commons and House of Lords Select Committees on Science and Technology, Parliament does take a detailed interest in the R&D spending of government departments (and has made severe criticisms at times) but their Reports, which are considered by Cabinet, have not been especially influential in the past. Through Parliamentary Questions, individual ministers can also be questioned on issues relating to R&D activities. Such parliamentary oversight acts in a rather general way, making departments and ministers aware of the need to act in a way that is generally consistent with the aims of Parliament, but parliamentary intervention seems to have little direct impact upon the activities and spending of individual ministries.

Consultation and Networking

The main formal advisory body involved in R&D in the UK is the Council for Science and Technology (CST). The CST (formerly ACOST, before that

ACARD) is an advisory committee with a wide-ranging remit to advise ministers collectively on the balance and direction of UK SET. The Council brings together senior representatives from business, academia and research, all appointed on personal title by the President of the Board of Trade. A balance in the representation of interests is aimed for across academia, industry and users of new technologies. Past items on the CST's agenda included: measures to enhance the contribution of SET efforts to national wealth creation and to the quality of life; implications of the government's health service reforms for research in the NHS; efforts by the MOD to transfer benefits of its R&D to the UK civil sector; and the DTI's strategy for promoting successful innovation in UK industry. The CST may advise government on involvement in the support of particular research programmes. In particular, it has occurred that in the past a case has been made for halting support to specific programmes (for example, the fast breeder reactor). The CST, however, is seen as a body set up by government to advise government rather than as a lobbying pressure group.

Pressure groups and lobbyists do attempt to influence R&D decisions and there are many routes by which pressure can be exerted. Through individual Members of Parliament in committee or through parliamentary questions particular items can be brought to the notice of individual ministers although, as already stated, this will affect the general environment rather than have a direct impact upon policy. Through the lobbying of individual ministers lobbying action may be more effective, for individual ministers do have considerable power.

In addition, there are various routes by which discussion and consultation take place among members of the policy making community, and between civil servants, representatives of industry and other groups, that may feed in to the policy making and spending process. It appears that many formal and informal links between ministries and the different stakeholders in the economy provide numerous routes for influence. In the process of policy development, for example, at the DTI, various people are consulted, but mostly on a case-by-case, grass-roots basis. There are no large consensus making bodies across the board and no formal consultancy processes either. Unions are not represented directly. The DTI has civil servants with certain industry responsibilities who build up a detailed knowledge of their industry and a network of industry contacts. These civil servants may act as a conduit for the expression of industry views. Lobbying, in the sense that individual companies take direct steps to influence the shape of R&D policy, is unusual. Outlets such as industry associations appear to be adequate channels to put such views across. It appears that in various ways the voice of industry and business and the voice of academia are heard quite often and quite loudly in the corridors of power, and that policy does react to these voices. Despite the recent change in political regime, however, the voice of organized labour is not very loud in discussions on this topic.

The UK government is attempting to strengthen co-ordination between research requirements for its own policy purposes and the technological needs of industry. This has been done by establishing consultation procedures and by developing systems to optimize the extent to which publicly funded SET aligns with the needs of the wealth creating sectors of the economy. Where research is funded for policy, statutory and regulatory reasons, mechanisms are put in place to ensure that the potential spin-off for wealth creation and improvements in the quality of life is addressed. Among specific initiatives taken to ameliorate government-industry interaction are:

- the MOD's Pathfinder scheme, aimed at improving military or cost effectiveness in the development of future products, both military and civil;
- the DoE's Partners in Technology scheme;
- the DTI's Managing in the 90s programme, and the Business Link centres which assist SMEs with technology-related problems;
- the MAFF's collaborative work on research programmes with levy boards, with food, fish farming and water industries, and under the LINK scheme;
- the NHS's National Industry Research Advisory Group which is examining better ways of achieving interaction between the NHS and the medical equipment industry;
- the Northern Ireland Office Technology Development and Networking Programmes.

Increasingly, industry-directed R&D policies are designed to be delivered at the local level and, to the extent that they consult, local authorities may shape the implementation of such policies. This consultation is not a process fully under the DTI's control, as local authorities may be of the 'wrong' political colour, which makes formal links difficult. This used to be a problem when the last Conservative government was in power, but it has been removed by the 1997 elections.

COMMUNICATION NETWORKS

An example of a body which facilitates networking is the Defence Scientific Advisory Council (DSAC) and its five supporting boards; it provides the Ministry of Defence (MOD) with an independent view of the department's R&D programme. The DSAC draws its members from both industry and academia.

In the Department of Trade and Industry (DTI), to promote 'partnership' the department works with and through other organizations, such as the Confederation of British Industry (CBI), Training and Enterprise Councils (TECs), Trade Associations, and Research and Technology Organizations (RTOs). The department has concluded concordats with all six Research Councils and interacts on a regular basis with other departments with a strong research interest, such as the MOD, the Department of Health (DH), the Ministry of Agriculture, Fisheries and Food (MAFF), the Department of Transport (DoT), merged with DETR in 1997, and the Department for Education and Employment (DfEE). Usually, however, such contacts will be informal rather than formal.

There is also a whole network of senior- and working-level bi- and trilateral contacts that go on within government. There are, for instance, contacts between the DTI and the MOD dealing with synergies in the development of automotive vehicles. In aerospace co-ordination is most elaborate. The DTI funds joint programmes with defence research agencies and there are committee structures on which the MOD and the DTI and some of the companies involved sit to decide the direction research should take. There is also a Committee of Chief Scientists of various departments which tends to deal with cross-departmental multilateral issues. Furthermore, there is a high-level group involving Chief Scientists of the OST, the MOD and the DTI which looks at dual use and synergy between defence and civil R&D.

2.3 PHILOSOPHIES

Three rationales currently appear to underlie the structure of R&D policy in the UK:

1. The UK, as a major trading nation, will only realize and maintain economic prosperity if the economy is more competitive. Technological change is seen as a major route by which competitiveness can be achieved and the technological performance of the economy needs improvement.
2. Within the UK there are many world-class companies. However, there is

a long tail of other companies that are not of similar standing: much can be achieved by improving the performance of such companies. The world-class companies at the same time should be provided with an environment that enables them to continue to succeed.

3. The UK has a very strong science base in its universities. The strength of this base does not, however, reflect in the competitiveness of its industry. There is thus a need to further strengthen links between science (the universities) and industry.

THE 1993 WHITE PAPER

In May 1993 the UK government published a White Paper on science and technology policy: *Realising our Potential: A Strategy for Science, Engineering and Technology* (HMSO, 1993). On the basis of a widespread consultative process, this document pinpoints main areas of weakness in the UK national system of innovation and outlines a number of policy initiatives designed to counter these problems.

The White Paper signals a 'widely perceived contrast between our excellence in science and technology and our relative weakness in exploiting them to economic advantage'. Furthermore, government involvement in science, engineering and technology (SET) is seen as an area of concern, due to a lack of clear policy objectives and weaknesses in management of public investments in SET and in mechanisms for implementing SET policy.

Another major area of concern is the management of careers in science and engineering. Overall, the White Paper advocates the view that the main problem facing the UK economy is not a lack of ability to generate new technologies; the size and quality of the science and engineering base are judged to be adequate. The main problem of the system is its failure effectively and successfully to exploit new ideas, whether produced in the UK or elsewhere. Effective exploitation, not only of new technologies but also of new markets, new forms of organization and new skills, is seen as crucial to promote the UK's competitive position on world markets and to improve the quality of life.

Because the main problem of UK technological progress is not believed to be innovative capacity itself but rather the ability to exploit, the proposed remedy is not increasing investment but improving co-ordination: 'it is the fundamental theme of this White Paper that a closer partnership and better diffusion of ideas between the science and engineering communities, industry, the financial sector and Government are needed'. Therefore the government intends to forge intimate links between technology producers (for example, universities) and users of technology (such as decision makers in commercial enterprises and government departments); to manage government's own research more effectively; and to increase public understanding and acceptance of the importance of science and technology.

Specific policy initiatives announced in the White Paper centre around improvements in communication, generation and dissemination of information, co-ordination of initiatives and monitoring of progress: 'The aim is to achieve a key cultural change: better communication, interaction and mutual understanding between the scientific community, industry and Government Departments.' In particular: 'government's use of funds ... will be made more explicit and open'; 'A Forward Look will be published each year'; 'Technology Foresight ... will be used to inform Government's decisions and priorities'; 'technology transfer will be developed to re-emphasise the importance of the interchange of ideas, skills, know-how and knowledge'; 'There will be easier access, especially for small and medium-sized firms, to the innovation support programmes'; 'There will be a new campaign to spread the understanding of science and technology in schools and amongst the public'. In addition to and in support of the above, the White Paper announced a number of organizational changes with respect to Research Councils, to co-ordination of cross-departmental SET issues, and to education in science and engineering.

Although these general viewpoints are more or less shared as a starting point by all parts of the UK government and civil service, there are differences in focus between the different players. Within the general philosophy summarized above there is a wide range of emphases that are reflected in different policy attitudes and different policy stances across ministers, ministries and over time.

The Treasury

The most consistent philosophy over time has been that of the Treasury. Its main concern has been to control government spending and to minimize the public sector borrowing requirement while at the same time seeking lower tax rates in the long run. The overall aim is a reduction of public expenditure as a share of GDP. A secondary and related concern has been to ensure the effectiveness of and economy in government spending. Public expenditure must genuinely contribute to the creation of wealth. This latter concern has led individual ministers to increase the efforts put into programme evaluation and also was a driving force behind the establishment of the OST.

The Treasury influence on R&D budgets is largely through the role it plays in setting individual department spending totals. It is primarily up to ministers and secretaries of state to determine how to deploy departmental budgets to fulfil departments' objectives. Once a department has its budget set, the Treasury does not seek to second-guess a secretary of state; departments set the priorities. Moreover, the Treasury lacks the resources and detailed technical knowledge to do so. Although it employs economists to assist in discussions with departments and to support arguments for and against the rationality of particular spending proposals, it does not employ scientists and technologists to deal with the SET budget.

The total spend of any department is more important to the Treasury than the pattern of that spending across individual budget lines. However, given the Treasury's continual desire to cut government spending, departments are aware that they must be capable of justifying any spending *a priori*, or be able to show *ex post* that a programme of spending was effective. The Treasury is interested in the background as to how different items of expenditure contribute to the overall objectives. Elements in the negotiations are: the output expected from the expenditure; its contribution to national wealth; and whether expenditure in the past produced the intended outputs. Departments take these matters into account as they attempt to institute new R&D programmes.

In looking at departmental spending the Treasury applies two clear rules: justification for a programme of spending can only be made upon the grounds of market failure; furthermore, the spending itself has to show additionality. This essentially means that free market forces would not bring about the desired objective, that government spending is not replacing private spending, and that government spending will achieve the objective. In theory, market failure is a clear concept: market failure occurs when free market forces will not bring about the welfare optimal outcome. In practical terms, however, it is difficult to apply. There are numerous reasons why market failure can occur and all the literature indicates that in the area of innovation market failure is endemic. It is often not easy to show that if government does not provide funding, the private sector will not fund either, or that expenditure on fundamental research contributes to national prosperity.

It is our view, based on the discussions held, that the Treasury takes a rather narrow view of market failure. For example, it is accepted that most basic research activity is subject to market failures and thus there is a good rationale for government spending in this area. Even though the existence of market failure is accepted, when it comes to justifying particular levels of expenditure there is much discussion: it may be accepted that basic research is subject to market failure without accepting that any particular science project merits government support or that a particular spending level is desirable. Moreover, when it comes to issues of risk (the question whether the private sector underinvests in R&D because of risk aversion or inability to diversify risk) or of short-termism (private sector underinvestment because of too high discount rates), Treasury attitudes are much less likely to consider these to be market failures.

The Treasury's emphasis on cost effectiveness is one of the reasons that *ex post* programme evaluation (done either in-house or by external consultants) has become more common. However, it is often difficult to attribute current commercial successes to research projects done many years earlier; evaluation here is very difficult.

Office of Science and Technology

As explained above, the OST has two basic functions: to oversee the totality of government R&D spending without executive authority, and also to act as the government department responsible for the science budget. In the former task the revealed activities of the OST reflect a philosophy that efficiency in R&D spending is paramount, that overlaps should be eradicated and that interdepartmental co-operation should occur where relevant. In past publications the OST has argued in favour of a strategic plan for the totality of government R&D but that has never come to fruition. The OST has also emphasized the importance of evaluation of past spending.

CO-ORDINATION AND COLLABORATION

One of the main objectives of SET policy announced in the 1993 White Paper was to improve the organization of the 'national system of innovation' through more effective co-ordination of initiatives and through promotion of collaborative networks. In this context the government has established mechanisms to co-ordinate science and technology activities relevant to a number of different departments. The responsibility for this has been allocated to the OST. The OST has distinguished the following circumstances which require collaboration:

- where a major policy statement has been made and there is a need for SET programmes to be co-ordinated so as to assist in successfully delivering the policy objectives (for example, sustainable development);
- where advances in technology, if effectively co-ordinated between departments, hold out the prospect of new policy options (for example, transport telematics);
- where there are particular policy needs which SET is not yet adequately addressing (for example, risk assessment and toxicology);
- where interdepartmental collaboration can lead to more effective use of scarce resources, avoid duplication of research and facilitate technology transfer (for example, civil–defence SET collaboration).

In addition to intensifying co-ordination of SET activities across departments, the government has taken steps towards more effective co-ordination between government departments and Research Councils. To this end concordats between Research Councils and government departments have been established. These are to facilitate the flow of research results from Research Councils to departments, as well as to ensure that Research Councils take account of the research requirements of departments and other users, especially industry.

Through the Chief Scientific Adviser the OST provides advice to the Cabinet Committee on Science and Technology on how the R&D numbers add up across departments and ministries. We were told that this advice has never been particularly effective. In terms of the figures in the Public Expenditure Survey, the advice of the Chief Scientific Adviser does not seem to have made a significant impact on either the volume of SET spending or on the way departmental ministers use their resources. Ministers are unlikely to give up their rights to set priorities between policy areas and to use instruments within each policy area as they think appropriate. However, given that the OST reports to the Cabinet and thus can, if necessary, effect changes in individual ministries through this route, the ministries need to be aware of the conclusions reached by the OST and do take them seriously.

The other side to the OST is its responsibility for the science budget. In line with the dynamic nature of the determination of the R&D budget, the OST philosophy with respect to this science budget has been changing over the last five years. Based on the view that the strength of the UK science base is not reflected in the strength of UK industry, there has been a steady move towards linking the science spend to wealth creation. Under a series of ministers, although the importance of basic research was still recognized, government policy began to place more emphasis on an allocation of science spending to meet user needs. This emphasis has been embodied in the Technology Foresight exercise, which was a detailed study of user requirements from the science base. The Technology Foresight exercise has produced a detailed listing of priorities that governs the allocation of government moneys to science in future years and is being used as a guide by the Research Councils. This move towards linking science spending to wealth creation has not been wholeheartedly welcomed by the academic and scientific community although there appears to be widespread industrial support for it.

From the viewpoint of the OST, market failure is an important legitimizing factor for government spending on science. Scientific research should be supported by government expenditure because there are important externalities. The aim of supporting research in universities is not just to obtain research results but also to maintain research capacity, keeping high-level researchers active within the UK, fostering spill-overs into postgraduate and undergraduate education, and thereby creating an infrastructure and climate which will be attractive to companies making investment decisions. Supporting high-level technological research contributes to the upkeep of the skills basis. Science policy is thus an integrated part of the system of education and training. At the high-technology level this system seems to function adequately; in the UK skill shortages tend to be signalled at the technician level rather than at the postdoctoral level. Weaknesses at the technician level fall within the remit of the Department for Education and Employment.

Ministry of Defence

If we turn now to the philosophies of individual ministries or departments we again see a certain heterogeneity. The Ministry of Defence, which is the department spending most on R&D, sees its objectives as ensuring the security of the United Kingdom and enhancing its defence capability; its direct expenditure upon R&D is for the development of new weapon systems (although there have been changes over the years whereby weapons contractors have to cover more of the R&D costs themselves with recompense coming through sales of the final product). Although the ministry welcomes civil spin-off from this military spending and has taken some steps to encourage it, it does not accept the stimulation of UK technological capabilities or of the competitiveness of UK industry as prime objectives. In this it differs from, for example, the US Department of Defense. Partly as a result, the various policies that have been tried to stimulate civil spin-off from defence R&D seem to have had limited success. It might thus be considered that the emphasis upon defence R&D in the UK is driven by factors quite separate from those relating to generating competitiveness in the UK economy.

Department of Trade and Industry

The other major R&D spending department is the Department of Trade and Industry. The DTI exhibits a rather different philosophy from the Ministry of Defence. Its prime objective is to promote the competitiveness of UK businesses in world markets. Policy is now principally concerned with promoting business best practice. Innovation is understood as the successful exploitation of new ideas. It is recognized, though, that there is a need for complementary assets for technology to be exploited. Programmes like 'Managing in the 90s', for example, look at business processes as a whole rather than technology as such. Social and environmental objectives which contribute to the quality of life are not a central focus, but are pursued as long as they go hand in hand with promoting competitiveness. For example, the DTI provides support for work on consumer safety, measurement standards and product performance standards. While measurement and product performance standards make an important contribution to competitiveness, they also contribute to the quality of life (for example, measurement of radiation emitted by medical equipment). However, much of the work in the areas listed here is required by legal statute. In the targeting of its support the UK seems to put the accent more explicitly on the competitiveness issue than, for example, Germany or Finland.

Present policy is the result of a historical process. The recent history of R&D policy formulation has witnessed two major shocks: one in 1988, when a more direct involvement of the Cabinet was introduced, and one in 1993, with the publication of the technology White Paper. Since 1988 there has been a broadening of perspective from simply reacting to market failure to actively promoting UK competitiveness. For much of the 1980s, policy proposals were brought forward, by industry or from within the department, and then a market failure test was applied. The focus on competitiveness is more proactive, gives more guidance to policy formulation, and has created a more integrated approach across sectors. The market failure test is still important, though: it must still be shown that there is a need for government action and that government action will be cost-effective.

MARKET FAILURE AND SHORT-TERMISM

Government action is required if there is market failure. The DTI distinguishes the following main sources of market failure which are likely to affect the innovative activities of firms: externalities and spill-overs; lack of appropriate information available to agents in a market; public goods; perceptions of risk and uncertainty; lack of competition; and barriers to market entry. The main market failures relevant to innovation are nearly all related in one way or another to a lack of information or a failure in the market for information: capital markets that do not have the information to assess risks of technologically innovative projects; firms that lack access to information on business best practice; lack of awareness of technologies; shortages of knowledge and skills.

Likewise, short-termism, an undue emphasis on or preoccupation with the short-term profitability of investment, is seen as a consequence of imperfect information. The long-term build-up of intangible assets therefore gets neglected. There is probably short-termism on both sides of capital markets, that is, in the City and on the side of business managers. In the eyes of the DTI, short-termism is not a problem of structure (for example, the predominantly arm's-length relationships between business and finance), but one of operation. There is a mismatch of perceptions on both sides of the market. The solution is sought in improving communication, getting industrialists to talk to the City, explaining the benefits of innovation, and trying to get a better match of understanding of risk and rewards, given technical uncertainty.

Over the last ten years the DTI has increasingly moved away from support of R&D at the near market end of the spectrum. It now sees its task as being much more to do with enabling technological development and removing constraints, rather than direct spending. The task of the DTI is seen principally as to create conditions, less so to identify direction, and to take specific initiatives only when justified, almost by exception. The DTI tries to moderate a process, to work with industry so as to extract views from industry and to act as a sounding board. Firms are primarily responsible for their own performance; the DTI's role is to help them to help themselves. To take the lead is generally not seen as the government's task. The DTI objective is

'Innovation in the UK', where innovation is defined as the successful exploitation of new ideas. This objective, they argue, is best served by encouraging best management practice (such as benchmarking), spreading knowledge, encouraging links between marketing and technology development, rather than subsidizing R&D spending itself. This approach is consistent with the view that the UK problem is one of bringing all companies into line with the best. Potential constraints such as short-termism are tackled through a belief in the efficacy of markets and policies designed to improve market performance.

Another important aspect of R&D policy has been the very welcoming stance taken by government with respect to overseas multinationals wishing to set up production and perhaps R&D facilities in the UK. Although this may partly reflect regional policy objectives, such new plants are seen as a main route by which new products and processes may be introduced into the UK, and also a means by which (world) best practice may be disseminated through UK companies, either as a result of competitive pressure or through the supply chain to component and other suppliers. This openness to overseas investment is also a reflection of the view that the primary role of the City of London in international capital markets should be reflected in the government attitude to overseas investment in the UK. If major technical and commercial opportunities are identified, but no facilities or abilities are locally available, it has happened that the UK government, through the DTI, has encouraged foreign companies to come in and fill the gap. There is now, for example, a sizeable TV industry in the UK which is not British owned.

In the DTI there is also great interest in support measures overseas and in competitive positions relative to foreign countries. Though support policies abroad do not themselves justify expenditure, what foreign governments do provides important signals. International comparisons are important and may trigger action and indicate direction. Matching foreign policy, using the argument of creating a 'level playing field', is an issue that is relevant to the aerospace industry, for example. The DTI has always felt that on occasions there was a need to match subsidies abroad, but the Treasury tended to resist this very hard; matching played a controversial role. Nevertheless, there is no reason to match R&D subsidies abroad, even in a dynamic context, if there is not some sort of payoff. In the case of the aircraft industry it would be considered that the industry creates extensive spin-off. Therefore, even if it is not exceptionally profitable, it is still valuable to the economy. There would be a cost to losing it and thus a case for matching subsidies (there would not be a case for creating this industry from scratch, though). Such arguments seem to carry increasing weight and to be gaining recognition in Treasury circles. Matching foreign policies is a reason for not doing less, but it is not a main imperative for doing more.

The power of individual ministers when considering departmental philosophies should not, however, be discounted. Under Lord Young the DTI specifically emphasized co-operative R&D and technological development. This emphasis has now all but disappeared. Under Michael Heseltine there was a more open attitude to intervention in industry, although in practical terms there is little evidence that this was put into effect. Under the three ministers who have held the post since Labour came to power in 1997 further reviews of practice have been taking place. For some time, though, there has been a shift of resources towards technology transfer and networking. However, the information travelling on the communication networks no longer emphasizes innovations *per se*, but rather softer information, for example, on best practice up and down the supply chain, on business development, and on bench-marking.

Throughout the last ten years, in line with a general government philosophy of privatization, there has been a gradual move away from R&D actually being performed in government-owned establishments. Institutions such as the National Physical Laboratory have been privatized and now only work for government through contracts. This has also led such institutions to seek private sector work to supplement the publicly funded work that they undertake. Complementary to this, universities in the UK have also been encouraged to increase their funding arising from the private sector and from competitive bids for government contracts, and have made remarkable moves in this direction.

2.4 POLICIES

What we have seen over the last ten to fifteen years is governments putting continual pressure on the totality of government spending and trying to make economies wherever possible. In terms of the R&D budget this has meant pressures to reduce the total budget and also an understandable desire for efficiency in the spending of that budget. We also see a belief in the efficacy of free market forces leading to the conclusion, first, that industry should itself pay for R&D (except where market failure exists), especially at the near market end of the spectrum, and second, that government research establishments should wherever possible be privatized. Government support for R&D moved away quite dramatically from experimental development towards basic and strategic research. This was partly a consequence of the decrease in spending on defence-related R&D, however. Also, more and more R&D support schemes are in the form of straight competitions where applications are invited. The number of programmes for which applications are made serially and the best are picked according to a set of criteria has been decreasing. In 1996–7, 62 per cent of government spending on SET was subject to competition. Half of the funds flowing through the Research Councils were competed for, 52 per cent of the SET expenditures of the civil departments and 76 per cent of the spending of the Ministry of Defence.

The changes in priorities and policies that have been discussed above can be summarized by data on changes in the socio-economic objectives of R&D spending in the UK over the 1986–7 to 1996–7 period (see Table 2.3). Whereas expenditure on defence fell from 45 per cent to 37 per cent of the total, on industrial development from 10.5 per cent to 2.5 per cent, and on energy from 4.4 per cent to 0.7 per cent, expenditure on advancement of knowledge rose from 21.5 per cent to 29.7 per cent, on health from 4.5 per cent to 14.5 per cent and on environmental protection from 1.1 per cent to 2.2 per cent.

Table 2.3   *Government funding of net R&D by socio-economic objective in 1996–7*

	Expenditures (£ million)	% share
Agriculture, forestry and fishing	259	4.5
Industrial development	144	2.5
Energy	40	0.7
Infrastructure	98	1.7
Environmental protection	127	2.2
Health	835	14.5
Social development and services	121	2.1
Earth and atmosphere	98	1.7
Advancement of knowledge	1711	29.7
Civil space	161	2.8
Defence	2142	37.2
Not elsewhere classified	23	0.4
Total	5759	100.0

Source: Office of Science and Technology, *Science, Engineering and Technology Statistics 1998*, OST Website (http://www.dti.gov.uk/ost/).

The Science Budget

After defence, the next most important area of spending is that which we can label the science budget (although there are several different ways of defining this term). The first point to note is that, unlike in many other countries, in the

UK nearly all basic research is undertaken in universities. Very little (although some) is undertaken in industry; but more to the point, in the UK freestanding research institutes do not have anything like the presence that we see in Germany, France, the Netherlands and Finland (although with their privatization some previous government research labs are beginning to look like such institutes). The universities combine basic research with their teaching function and increasingly are also becoming contract research institutions, undertaking applied research and development for the private sector. Until a few years ago there was a binary divide in the UK tertiary education system between universities and polytechnics. All polytechnics have now been labelled universities.

HIGHER EDUCATION FUNDING COUNCILS (HEFCs)

There are separate HEFCs for England, Wales and Scotland, the members of which are government, university and industry representatives chosen by the Secretary of State for Education and Employment. The HEFCs allocate a block grant to each university to fund teaching activities and also to provide basic research facilities (the concept of the 'well-found laboratory'). The allocation of such moneys is dependent upon regular reviews of teaching and research performance in different universities: the higher the rating, the higher the allocation. The HEFCs in conjunction with the relevant Secretary of State decide on a split within the overall budget for universities between funding for teaching and funding for research. This is a notional split only; it is for individual vice-chancellors to decide how they allocate funds in practice. This funding is supplemented by a subsidy per student paid by the local authority of the area in which the student normally resides (in the UK most students attend a university away from their home) and for postgraduates paid by the student. In the HEFC allocation there are attempts to provide more undergraduate places in areas of perceived shortage (such as science and technology) but no detailed manpower planning takes place. Over the last decade the proportion of the age cohort attending university has increased from 20 per cent to 33 per cent (compared to 3 per cent in the early 1960s).

Universities are partly financed by government and partly (and increasingly) through contract research and teaching for other bodies, although compared to US universities the share of industry finance is still very small. Government finance comes through two routes. The first is from the Department for Education (and now Employment) and is distributed via the Higher Education Funding Councils (HEFCs) and provides the means for teaching and basic research (£1.012 billion in 1996–7). The second route by which the universities are financed is through the Research Councils (£670 million in 1996–7). The Office of Science and Technology argues over the total science budget with the Treasury. That budget is then largely allocated to the Research Councils for distribution. After a reorganization at the beginning of the 1990s there are now six Research Councils. In order of size they are: Engineering and Physics; Medical; Particle Physics and Astronomy; Biotechnology and Biological; Natural Environment; and Economic and Social. The Councils are overseen by the Director General of the Research Councils who reports directly to the President of the Board of Trade. Although the Director General does not have a standing Advisory Committee like the former Advisory Board for the Research Councils, which he replaced, he does consult universities, research council institutes, industry and various industrial groupings to obtain advice. The membership of the Research Councils themselves reflect academia, government and industry. The Councils largely allocate their funding on the basis of excellence based upon peer review of researcher-originated bids for funding (responsive mode funding). However, although the principle of excellence still applies, a policy stance which relates science spending to wealth creation and the needs of UK industry is currently being put in place. This has been driven by the Technology Foresight exercise which is to be continually updated as time goes on.

An advisory committee, the Science and Engineering Base Co-ordinating Committee (SEBCC), chaired by the Chief Scientific Adviser and consisting of chief executives of Research Councils and Higher Education Funding Councils and of representatives of the education departments, is charged with the task of providing effective working links between the Research Councils and the HEFCs, in order to achieve an effective, mutually reinforcing partnership between the two sides of the dual support system.

A detailed review of the science budget portfolio in 1995 has led to a major reorientation of the composition of the portfolio towards the White Paper objectives (Cabinet Office, 1995). The Director General of the Research Councils concluded that the UK has a world-class science base. However, excellence is not spread evenly; within this resource there are stronger and weaker research groups:

> Many, perhaps most, of the science and engineering areas in which the UK has a world class reputation are those where there is:

(i)   adequate funding over a number of years;
(ii)   access to the often expensive equipment necessary to conduct world-class research;
(iii)   good interaction with industry and commerce, especially at the generic or strategic level, rather than solely based on contract research projects;
(iv)   a good supply of top quality motivated and highly trained researchers with a reasonably evenly spread age structure ...'

The analysis led to the targeting of three main areas. First, initiatives were taken to optimize interaction between academe and industry. The Research Councils have been improving their mechanisms for relating to their user communities. Also, the Realising Our Potential Award (ROPA) scheme, which rewards researchers who have gained industrial funding for strategic research by enabling them to pursue a curiosity-driven topic of their own choosing, has been extended. Second, enhancements were made to the funding mechanisms of basic and strategic science. The need to strengthen responsive mode funding in key strategic areas was recognized in the budget allocations. Peer review remains the key mechanism used by Research Councils to allocate research grants. Third, enhancements to people-related programmes have been announced. Arrangements are made to improve career opportunities of research staff and to provide extra support to promising PhD students.

Total government funding of higher education and science increased by 12 per cent from £2.089 billion in 1986–7 to £2.340 billion in 1996–7, in real terms, and from 29 per cent to 40 per cent of total government funding for SET (excluding the NHS) over the whole period. Of this total, the government science budget increased by 39 per cent from £946 million in 1986–7, in real terms, to £1312 million in 1996–7, but over the same period the funds distributed through the HEFCs decreased by 10 per cent from £1143 million to £1028 million. This shift reflects the drive to allocate more of the funds through competitive allocation mechanisms and to increase the responsiveness of funding to user needs. In real terms, both the science budget and the higher education budget are planned to go down up to the year 2000 (DTI, 1997a, 1997b).

Support for Industrial R&D

The main department responsible for the support of industrial R&D is the Department of Trade and Industry (the DTI, which also incorporates the responsibilities of the old Department of Energy). In the DTI we see especially that support for expenditures on R&D has been cut back considerably. Earlier schemes of government contributions to product and process development in private industry have been replaced with schemes that emphasize 'innovation' by encouraging (managerial) skill development, good managerial practice and that enable changes to the market environment. There is now more emphasis upon demonstrator schemes, for example, and also upon publicizing best practice in good companies in order to encourage emulation. There has also been a shift over the years away from support of large companies towards support for SMEs, and to the extent that government funding for technological development is available it tends to be concentrated on the SME end of the spectrum.

LARGE AND SMALL FIRMS

Present Department of Trade and Industry (DTI) spending policy puts a heavy emphasis on small and medium enterprises (SMEs). The rationale for this is predominantly pragmatic: financial help makes more of a difference if given to SMEs rather than to big firms, which spend much on R&D already. Also, small firms are more affected by market failure: their in-house capabilities are weaker, their managerial resources less developed, and therefore they are easier to help. Many schemes are predicated on the fact that firms need to adapt technology for their own use but lack the in-house capability to do so. A number of programmes, such as the Teaching Company Scheme, give people in companies better tools to work with, better perceptions. There is the hypothesis that SMEs are a source of growth, and that the British economy lacks a sufficient number of SMEs. The DTI's view on this issue would be that, given the role likely to be played by large UK-based firms, encouragement for innovative and fast-growing SMEs must be provided if the UK is to take sufficient advantage of the opportunities to exploit new technology which may come about.

Policies that meet the needs of large firms best often do not involve direct money hand-outs from government. Their problems have more to do with effective management, internal bureaucracy and communication. Financial support to large firms seemed to have leverage in the early 1980s, after a period of financial crisis, but since then the additionality of support to large firms has become questionable. Assistance for market research has gradually been scrapped. Current emphasis regarding large firms is on their relations with their environment and with the City, encouraging benchmarking, and changing the mind sets of management and principal shareholders. Sometimes large companies

> are involved in collaborative programmes supported by the DTI.
> There may be an interest in their involvement to the benefit of
> others, for example to act as a test bed for the use of a new
> technology. Nevertheless, the general principle is that they should
> finance their own investments as much as possible.

In real terms the DTI spend on R&D has fallen by 40 per cent from £581 million in 1986–7 (excluding OST and Launch Aid) to £356 million in 1996–7 and further reductions have since occurred. This is a clear reflection of how the government has pulled back from direct support of R&D in industry. It should be noted that a large part of the DTI R&D spend still goes to aerospace and nuclear energy. (Traditionally the UK government R&D spend has been orientated towards support for electronics, nuclear energy and aerospace.) On the other hand, the DTI spend (in constant prices) on technology transfer activities has reduced by less (one-third) between 1986–7 and 1996–7, reflecting the reorientation discussed above. We could characterize the present situation as one where DTI policy is orientated more towards the diffusion of technology than towards R&D. The DTI itself characterizes its current policy stance as one where what is emphasized it is not the technology part of innovation but good management practice leading to the successful exploitation of technology.

In practice this policy implies stimulating discussion, making contacts and networking. The DTI takes initiatives in getting other people to take a lead and in getting senior industrialists to talk under their own banner. The R&D Scoreboard, for example, is a way of raising consciousness. The DTI stimulates benchmarking to make businesses think about their performance. Development of accounting standards to reflect the benefits to innovation and publication of R&D figures also help to focus on the longer term. If accountants' information clearly showed the long-term value of innovation, then the long term would carry more weight. These are forms of indirect support to innovation where it is felt that great leverage could be obtained. Potentially large flows of money are involved, for example, through pension funds, if these initiatives are effective.

In line with the policy of 'enablement' rather than of subsidization, the DTI has also been encouraging activities that might improve the effectiveness of markets, for example, exhortations against short-termism in capital markets. These in turn are taking place in an environment where widespread privatization of previously government-owned utilities has been taking place. Such privatization has had an impact on R&D in more than one way. Research outlays in the privatized utilities (but also in British Telecom and in coal mining) have become much smaller than they were and research has become much more applied. After privatization these industries became more narrowly focused. Also, one thing that affects innovation is the willingness of domestic customers and firms to try out new things; innovation needs feedback. Nationalized industries in the past provided test beds for other companies: the British mining equipment industries, for example, needed reference sites, which the Coal Board used to provide. Privatized industries are less prepared to try out things. This type of supplier–user relationship was especially important to firms serving nationalized industries, such as those providing railway equipment and mining equipment.

LINKING SCIENCE AND TECHNOLOGY

Science and technology policy are complementary. The Department of Trade and Industry (DTI) therefore has links to the policy circles which determine science policy. The DTI, as well as industry, is represented on most of the Research Council boards. It has contacts with the Higher Education Funding Councils, and both directly and through industry links tries to influence how funds are directed and dispersed. Furthermore, there are practical links through the LINK programme and the Teaching Company Scheme. The Technology Foresight programme has produced some consensus views on where future markets may match future science and technologies. This has created some common vision in government, industry and academia, the main stakeholders in science funding, about where science funding could complement industrial research policy. Thus, where DTI finances research, for example through LINK, those LINK programmes are chosen on the basis of the results of the Technology Foresight programme. Where Technology Foresight has produced recommendations regarding technology transfer and infrastructural issues (financing, skills, and so on), the DTI can take action, for example, through regulation and development of the communication infrastructure.

To some extent the generation of initiatives is *ad hoc*. When at some instance, such as in the context of the Technology Foresight programme, a promising opportunity for the development of a new technology or market is identified, the DTI sees it as its task to moderate action. It brings together players who are likely to be interested and tries to co-ordinate. This happens regularly; for example, in the case of the use of gallium arsenide, where a collaborative programme was initiated; and in the cases of opto-electronics, robotics and so on. Sometimes these programmes break down: the gallium arsenide programme faltered because one of the firms withdrew, due to differences of view between senior management and the R&D department. Some lines of research in the Joint Opto-electronics Research Programme were not taken up by UK firms because early market demand turned up in civil segments of the market whereas the main participating firms were significant incumbents in the military market. The robotics initiative was successful, as was the organization of 'clubs', such as the 'time to market club' and the 'M62 club' dealing with sensor technology.

Technology policy at the DTI is not directed at steering the structure of the UK economy. Nevertheless, policy intervention depends upon the type of sector. Intervention in, for instance, the chemicals industry takes a different form than in the furniture industry; the former is highly organized whereas the latter knows hardly any form of organization. Compared to the situation in Germany, for example, where there is a strong institutional environment, the British situation shows structural weaknesses. This leads to the DTI doing things that otherwise industry associations could do.

Public Procurement

The whole question of whether government procurement should be seen as a tool of technology policy seems to be open to some confusion. Although this has been recommended by the House of Lords Select Committee and others, the Treasury indicated that procurement should be as economic as possible. Public procurement is not seen by the Treasury as an instrument of R&D policy. The guiding principle is that government buys value for money (which tends to be narrowly defined). Procurement is not to be used to achieve particular economic objectives, such as to promote or protect domestic industries, and spin-offs are not a valid reason for paying more for a product that can be acquired more cheaply in the absence of spin-offs. The DTI, however, see public procurement as a potentially strong instrument of technology policy in some fields, not only in military technology, but also in pharmaceuticals where the NHS is a major customer. Procurement policy in the NHS has in fact been used in a way that helps pharmaceutical companies to succeed in the UK. However, the difficult question is whether this instrument can successfully be used to stimulate technological progress without compromising too much its primary objectives, such as value for money. Lately, though, procurement decisions by spending departments tend to be taken with their industrial implications in mind. With the privatization of nationalized industries, however, opportunities to use procurement policy have diminished.

2.5 POLICIES TOWARDS EUROPE

The *Forward Look* (Cabinet Office, 1994) states that 'the UK government will play an active part in developing arrangements designed to ensure that, reflecting the principle of subsidiarity, the EU's research effort is focused on activities which add genuine value over and above national efforts'. The UK recognizes the value of collaborative research with its EU partners and regards EU R&D programmes as a valuable complement to its own programmes. However, the prevailing political philosophy in the UK does not show wholehearted commitment to Europe, irrespective of whether the Conservatives or Labour are in power. There is an undercurrent in UK attitudes that expresses a fear that the economic policies and liberalist attitudes which have been instituted in the UK over the last two decades are under threat from Europe.

Structures

The OST co-ordinates UK interest in EU R&D issues. In the run-up to the Fifth Framework Programme, OST is working in collaboration with other government departments with SET responsibilities to identify the UK's priorities for the programme. Through a process of consultation with departments, industry, the Research Councils and a wide range of representative bodies, a balanced set of objectives is defined which focuses on the ways in which EU collaborative research can best complement UK efforts and contribute to the goals of improving competitiveness and the quality of life. These objectives are pursued throughout the negotiations.

Framework programmes are agreed unanimously in the Council of Research Ministers, where the OST represents the UK. Individual programmes are then agreed (by qualified majority) with UK interests again represented by the OST acting on a departmental brief. The Treasury, however, represents the UK in the Budget Council where the year-by-year financing of programmes is agreed (and where there is qualified majority voting and large-scale trading between blocks of budget lines). Here research and technology funding is always a minor budget line compared to structural or agricultural funds. The OST and other government departments participate in the EC Programme Management Committees which supervise particular programmes in the Framework programme. Thus the UK government representation in Europe is quite widely based but mainly located in the Treasury and the OST.

OST efforts are directed at seeking to ensure that the programme is managed as effectively as possible and encouraging a high level of UK participation in Framework IV. The UK Research Councils maintain a joint office in Brussels to assist in the search for funding by academic institutions in the

UK. There are also a number of private sector consultancy firms that will assist organizations seeking funding from Brussels.

Budgets

The UK contributed £373 million to the EU R&D budget in 1996–7, up 147 per cent in real terms since 1986–7. The main expenditures for the Research Councils are contributions to the European Space Agency (ESA) and to the European Laboratory for Particle Physics (CERN). The UK has, together with the German government, taken strong action to contain costs at CERN, which have been escalating. The costs of ESA are also still a cause for concern.

The relationship between the spending of the UK government and the spending by the EU on R&D programmes is by no means clear. Formally the UK government stance is designed to produce complete non-additionality in EU spending (EuroPES). The cost of UK contributions to EU research pro- grammes are included in the overall control totals of public expenditures and considered in each year's public expenditure survey. Any increase in the contribution to the EU budget over and above the 1984 contribution is attri- buted to the total spending of individual departments by a mechanism that is not totally transparent and forms part of the total spend of that department. An increase in the UK contribution to EU research programmes would thus appear as an increase in the spending of an individual ministry or department and, as such, unless the spending total of that ministry or department is allowed to increase, there would need to be an equivalent reduction in other spending (which in principle could be non-R&D spending).

Clearly, if the government spending total were never allowed to increase, this route would ensure that, as far as total government spending is concerned at least, there would be complete non-additionality, that is, any extra contri- bution to European programmes would mean an equivalent reduction in UK spending. However, it should be noted that the attributions are on contribu- tions and not on receipts. Thus, to the extent to which the UK is successful in obtaining grants and contracts under, for example, the Framework programmes that exceed the contributions, the EU spend is completely additional to the UK domestic spend.

In practice, however, with respect to contributions, the degree of non- additionality depends upon whether, in the light of greater attribution to the budget of a department resulting from increased contributions to the EU, that department is able to argue for an increase in its spending totals. This de- pends upon the bilateral discussions that take place between ministers, their civil servants, the Treasury and the Cabinet each year over departmental spending totals. Departments have frequently been able to secure compensation from the Treasury to offset budget cuts that might have resulted from this system. Overall, however, we are unable to measure the actual spending impact of this system of attribution.

What is clear, however, is that to gain compensation departments would have to argue that the European programmes do not coincide with domestic programmes, priorities or objectives, and thus that domestic programmes are additional to European programmes. This ensures that EU programmes are not seen by departments as a free good and that EU programmes are taken into account when domestic programmes are designed and implemented. It may be argued that as departments have a strong incentive to design their programmes as complementary to the European programmes, the mechanism induces a process of careful co-ordination of positions taken in Brussels, where representatives from the UK contribute to the shaping of European initiatives, and in London, where the government departments design their own programmes. If precompetitive, collaborative, cross-border R&D is in accordance with a department's policy objectives and if it can be arranged through Brussels, then that option is preferred. For example, the reason why the Alvey programme (for IT R&D) was not followed up was that so much was being done through ESPRIT (European Strategic Programme for Research in Information Technology); the reason why the DTI has a low spending on telecommunications R&D is that so much funding has been going through the RACE (Research and Development in Advanced Communication Technologies for Europe) programme.

2.6 CONCLUSIONS

UK R&D policy has been changing in terms of both structures and instruments. This has been taking place in an environment that stresses market efficiency, privatization and reduced government spending. Actual policies are conditioned by consultation processes but these are often informal rather than formal. Parliamentary control is general rather than specific. The outcome has been reduced government R&D spending, a considerable reduction in support for industrial R&D, but an increasing emphasis upon the advancement of knowledge. Defence R&D has continued to be pre-eminent in the government R&D spend, but its share is declining. Expenditure on knowledge advancement is now being more closely tied to wealth creation and user needs through the Technology Foresight exercise. The need for improvement in technological performance in the UK is as strong now as it has ever been and to date it is an open question whether current policy meets those needs.

In July 1998 the new UK Labour government announced the results of its comprehensive spending review. In addition to extra expenditure on education and health over the following three years, an increase of £1.1 billion for science and research was also announced. However, some £0.6 billion of this funding was to come from a private charity, the Wellcome Foundation, reflecting a closer link between private and government funding than has been the case in the past. In addition, the announcement heralded two further changes in the fund allocation process in the UK. First, there has been a move away from annual spending budgets: budgets in this latest exercise are determined for three years rather than one. Second, there are suggestions that the Treasury will in the future monitor and examine the spending activities of departments in much more detail than has been the case in the past. It is worth noting that the immediate reactions of commentators centred upon whether the three-year budget system could be maintained when revenues were not known three years in advance. The greater involvement of the Treasury in departments was seen as an extension of the power of the Chancellor relative to that of individual ministers. As a final postscript we note that in late 1998 the DTI published a new competitiveness White Paper. The core of this document laid out a vision for the UK as a knowledge-based economy and is intended to initiate a discussion of how such a vision can be realized.

REFERENCES

Barber, J. and G. White, 1987, 'Current policy practice and problems from a UK perspective', in P. Dasgupta and P. Stoneman (eds), *Economic Policy and Technological Performance*, Cambridge: Cambridge University Press.

Cabinet Office, Office of Public Service and Science, Office of Science and Technology, 1994, *Forward Look of Government-funded Science, Engineering and Technology*, London: HMSO.

Cabinet Office, Office of Public Service and Science, Office of Science and Technology, 1995, *Forward Look of Government-funded Science, Engineering and Technology*, London: HMSO.

Cabinet Office, Office of Public Service and Science, Office of Science and Technology, The Director General of Research Councils, 1995, *Review of the Science Budget Portfolio*, London: HMSO.

Department of Trade and Industry, Office of Science and Technology, 1996, *Foresight Programme: First Progress Report 1996*, London: HMSO.

Department of Trade and Industry, Office of Science and Technology, 1997a, *Allocation of the Science Budget 1997–1998*, London: HMSO.

Department of Trade and Industry, Office of Science and Technology, 1997b, *SET Statistics 1997*, London: HMSO.

Department of Trade and Industry, Office of Science and Technology, Chief Scientific Adviser, 1997c, *The Quality of the UK Science Base*, London: HMSO.

Department of Trade and Industry, 1997, *The UK R&D Scoreboard 1997*, URN 97/31, DTI London in association with Company Reporting Ltd.

EZ, 1997, *Toets op het concurrentievermogen 1997: Klaar voor de toekomst?*, Den Haag.

HMSO, 1993, *Realising Our Potential: A Strategy for Science, Engineering and Technology*, White Paper, London: HMSO Cm2250.

OECD, December 1997, *Main Science and Technology Indicators*, Paris: OECD.

Technology Foresight Steering Group, 1995, *Progress Through Partnership*, London: HMSO.

Walker, William, 1993, 'National innovation systems: Britain', in Richard R. Nelson (ed.), *National Innovation Systems: A Comparative Analysis*, Oxford: Oxford University Press.

3. The Netherlands

3.1 INTRODUCTION

The Netherlands is a small European country located between Germany, France and the UK. Although geographically not entirely correct, this observation holds some ground as a description of the Dutch political and administrative culture. The fact that the Netherlands is a small country with an open economy seems a factor constantly on the minds of Dutch policy makers and is a frequently recurring theme in policy discussions. It implies a high interest in maintaining access to international markets and 'level playing fields'. It also implies a high degree of dependence upon foreign sources of technology and foreign willingness to invest in the domestic economy. This dependence makes the Netherlands acutely vulnerable to international competition and makes its policy makers very sensitive to increasing trends towards internationalization. Virtually every policy document on Dutch technology policy contains an extensive international comparison and an analysis of the consequences of 'globalization'.

The Dutch administrative culture combines elements of the German corporatist culture, French étatism, and British liberal tradition. Liberalism, both as a guide for policy making and as the basic legitimation of market allocation, is an endemic element in the organization of Dutch society and is still gaining ground. Since the seventeenth century, the Netherlands has thrived as a trading nation. Trade and distribution are considered fields where the Netherlands has a comparative advantage; the country has never been particularly focused on manufacturing. The openness of the Dutch economy and its dependence upon trade has traditionally forced the Netherlands to take a pro-market stance in matters of international relations and trade; it is felt that any protectionism is likely to backfire. There is a worldwide tendency to use the market increasingly as the preferred mechanism for the allocation of resources. This trend is related both to technological developments that increase the feasibility and efficiency of market allocation and to cultural developments which lead to a shift from collectivism toward individualism. The Netherlands considers itself to be in a good position to take advantage of the liberalization of global markets.

Étatism is not an overriding trait of the Dutch administrative tradition but it is apparent at some points. The relationship between government and the knowledge infrastructure is characterized by a heavy state involvement, not only through the determination of budgets but also via the determination of research priorities and dissemination procedures. Dutch universities have limited autonomy in their policies on the establishment of departments, tuition fees, entrance requirements and salaries. Many of the major research facilities and laboratories in the Netherlands are either state institutes or former state institutes which retain a heavy state involvement.

The system that has produced one of the success stories of Dutch technological prowess can serve as an illustration of recent developments. In agribusiness, especially in primary production, innovation was organized predominantly as a top-down process. A government-maintained centralized system produced innovations in agricultural products and processes. Through a hierarchical system of laboratories, testing stations and experimental farms, innovations were developed step by step for commercial application and introduced to farmers for use in production on a large scale. Under pressure from changed circumstances these arrangements are currently gradually being transformed. Markets nowadays pay a premium for a more differentiated range of high-quality products. Innovations to exploit these new market demands are originated bottom-up rather than top-down. Recognizing this, the government is now withdrawing step by step from steering the agricultural knowledge system top-down. State research outlays are being put at arm's length and the system is being transformed into a market-driven organization (with the government still by far the largest customer). At the same time the government has increased its efforts to support and subsidize entrepreneurial farmers in generating innovations bottom-up and thereby exerting demands on the knowledge infrastructure. For the research institutes this comes down to a change from direct government support to indirect support by way of the users of knowledge.

Corporatism as an administrative mechanism is maybe most typical of the Netherlands. Its policy making process, as well as its industrial relations and its mechanisms for conflict resolution, is mainly corporatist in spirit. There is a constant search for consensus among various interests. Policy making therefore usually involves extensive consultations and rounds of bargaining with all the parties involved or affected. Numerous councils and committees bring representatives of interests together in discussion fora. Creating widespread support among stakeholders as part of the decision making process is seen as very important to ensure the workability and practicality of policies. The relatively small size of the Netherlands, where the number of players is limited and 'everybody knows everybody', makes such a procedure viable. This holds especially in rather technical and specialist fields like science and technology policy. Policy makers readily recognize that this system makes policy development sometimes very slow and radical changes in policy orientation quite unlikely. Also, vested interests are usually protected from, or compensated for, too-harsh consequences of policy changes. All this is seen at the same time as both a strength and a weakness of the Dutch system. A lack of decisiveness and versatility and a reliance on compromise are balanced by a high degree of stability and broad-based agreement on policy.

The origins of the corporatist tradition go back to the formative days of the Dutch parliamentary democracy. The Netherlands was a society composed of a number of, until recently fairly isolated, social blocks ('pillars'), most of them defined in religious terms and each a minority by itself. The largest blocks were Protestants of various types, Catholics and Socialists. This structure was vertically segregated in the sense that it ran across economic classes. Every pillar had its own middle class and labour class, its own social and political organizations, trade unions, media, and so on. Such a structure necessitated policy mechanisms at the state level that could bridge the differences between these communities and produce collective decisions. Every Dutch government was therefore a coalition government representing a number of social blocks.

As secularization, emancipation and the spread of higher education detached the individual over the years from the protective collective structures, the remnants of this old social structure have now nearly disappeared. The breakdown of the old social blocks is exemplified most clearly by the fusion of the large Protestant and Catholic political parties into a Christian Democrat party at the end of the 1970s, followed by the gradual marginalization of this party in the 1990s and the takeover of the political middle ground by the Socialists. For the first time since the beginning of the century, the Netherlands in 1994 elected a government in which no party with a religious background took part; Socialists came to govern together with Liberals. However, in spite of the dissipation of the traditional texture of Dutch society, the consensus system has remained strongly embedded. The economic successes of the mid-1990s (above-average economic growth, decreasing unemployment, low inflation, moderate wage growth and a decreasing government deficit) are often attributed to the Dutch version of consensus politics, referred to as the 'poldermodel'. Nevertheless, efforts are being made to increase transparency in the Dutch administrative system, with more stress on accountability, to increase the efficiency of policy development, to reduce the number of councils and committees and to alter the system of budget negotiations.

3.2 STRUCTURES

In the Netherlands, the national government accounts for about 41 per cent of spending on research and development. Dutch public policy on science, research and technology is mainly a matter of the Ministry of Education,

Culture and Sciences and the Ministry of Economic Affairs (see Table 3.1). Traditionally, agricultural research is an important separate cluster in the Netherlands. Dutch defence R&D is of comparatively little significance.

Table 3.1   Total government R&D funding, by department

	1997		2002 (planned)	
Department	Expenditures (millions GLD)	% share	Expenditures (millions GLD)	% share
Education (OC&W)	3360	62.2	3506	62.8
Economic Affairs (EZ)	912	16.9	1023	18.3
Agriculture (LNV)	352	6.5	325	5.8
Transport (V&W)	208	3.9	166	3.0
Defence	180	3.3	163	2.9
Foreign Affairs	163	3.0	162	2.9
Housing (VROM)	96	1.8	97	1.7
Health (WVS)	84	1.6	100	1.8
Other	45	0.8	45	0.8
Total	5400	100	5586	100

Sources: OC&W [Ministry of Education, Culture and Science]: *Onderzoek in cijfers 1995* and *Science Budget 1995.*

Table 3.2   Financial flows in the Dutch science and technology base, 1995 (millions of guilders)

Destination	Universities	Research institutes	Business firms	Foreign destinations	Total expenditures for outsourcing
Source:					
Government	2955	2207	274	—	5436
Business firms	153	435	242	401	1231
Universities	—	22	—	—	22
Research institutes	472	507	180	18	1177
Foreign governmental sources	131	78	94	—	
Other foreign sources	—	122	810	—	
Total net effort	3689	2707	6855		

Source: CBS, *Kennis en Economie 1997.*

Table 3.2 gives an impression of the flows of funds within the Dutch science and technology system.

Ministry of Education, Culture and Science

The main players in the R&D process in the Netherlands are the Ministry of Education, Culture and Science (OC&W) and the Ministry of Economic Affairs (EZ). Every ministry in the Dutch government has its own budget for science and technology (S&T), but more than 60 per cent of total government S&T expenditure is included in the budget of the Ministry of Education, Culture and Science. The Dutch scientific knowledge infrastructure includes the universities (nine general, three technical and one agricultural), the Research Council (NWO), the Royal Dutch Academy of Sciences (KNAW), the Royal Library (KB), the Netherlands Organization for Applied Scientific Research (TNO), the Large Technological Institutes (GTIs), and a number of national and international research organizations. The main part of this falls under the authority of OC&W.

The Minister of Education, Culture and Sciences is not only responsible for the largest share of total government spending on S&T, he or she is also the co-ordinating minister for science policy. There are no instruments of coercion at the minister's disposal to direct the science spending of other ministries; the function is solely of a procedural nature. The minister sets out, in consultation with colleagues, the main lines of government support for scientific development and research activities toward the basic end of the spectrum: basic, strategic and some applied research. Once every two years the OC&W publishes a document, the science budget, which gives an overview of policy intentions regarding the science policy of various departments. This budget is prepared at official level by representatives of the various departments most closely involved. It passes through different hierarchical levels until it reaches the Council of Ministers. Then the proposed budget is submitted to Parliament, is put on the agenda of Select Committees of both houses and, sometimes after amendments, is agreed.

Ministry of Economic Affairs

The Ministry of Economic Affairs (EZ) now has a much smaller share of the total government S&T budget than it had in 1985. It reached a maximum of 27 per cent in 1990 and has been decreasing rapidly to 17 per cent since then. The 1990 budget of 1294 million Dutch guilders has decreased to less than two-thirds. This decrease in the R&D budget reflects a decrease in direct R&D support to private business through subsidies. This trend parallels developments in the UK. It has a number of R&D support mechanisms aimed at promoting research in companies through development credit and R&D tax facilities (TOK and WBSO) and through grants for co-operative feasibility studies, research and demonstration projects (BTS). Major expenses are also its support of the microelectronics stimulation programme, the 'electronic highway programme' and the 'economy, ecology and technology programme', the European space exploration programme and the Dutch energy research organization (ECN). The Minister of Economic Affairs has a co-ordinating role across ministries with respect to technology policy, which includes activities ranging from applied research to development work. Also, EZ is the government's front door to the business sector and acts as an intermediary between the private and public sector. Large firms have ready access to EZ and they are often invited to voice their opinions. Captains of industry are asked for advice or are involved in advisory councils, either *ex officio* or as private persons. All the main independent advisory councils so far that have conducted inquiries for the government on the direction R&D policy should take have been chaired by prominent industrialists. Small and medium enterprises (SMEs) are far less involved in the policy development process.

Ministry of Agriculture, Nature Conservation and Fisheries

A third main spender on S&T is the Ministry of Agriculture, Nature Conservation and Fisheries (LNV) with a spend of around 7 per cent of the total government S&T budget (excluding its contribution to the agricultural university). Agriculture is a sector with heavy state involvement in every respect, ranging from market regulation to R&D support. The sector has a specific structure: it consists of more than 100 000 small enterprises, the majority of which are family businesses. Most agricultural firms are users of technology developed elsewhere. Because of these industry characteristics, responsibility for and co-ordination of technological development has largely been considered a task for the state. The network that has formed to ensure that this task is effectively performed in the interest of the agricultural sector is probably one of the most complex and tightly knit anywhere. Agricultural research is at the moment being concentrated within one large organization, the Wageningen University and Research Centre, spanning the whole range from fundamental research, performed at Wageningen Agricultural University (LUW), via strategic and applied research performed in 12 institutes that together form the Agricultural Research Department (DLO), all the way to testing in agricultural experiment stations. LNV finances more than half the costs of the LUW, about three-quarters of the costs of DLO, and provides 50 per cent of the funds of the experimental stations.

DLO was part of LNV until 1998 but has been made administratively independent (though not privatized). LNV continues to have a decisive influence on the design of research programmes, however. The determination of research programmes starts with the development of a policy proposal by LNV. Input is gathered from the research communities, prospective users and the National Council for Agricultural Research (NRLO). Reactions are invited from all sorts of stakeholders, most importantly from the Confederation of Dutch Agricultural and Horticultural Enterprises (LTO) and the various commodity boards (*Produktschappen*). These boards are organizations established by law, financed by industry and representing its interests. They operate on a closed-shop basis, have a number of regulatory tasks and co-finance research. The policy is finally submitted to Parliament as part of the budget.

Other Players

All other ministries are minor spenders; the Ministry of Transport, Public Works and Water Management, the Ministry of Foreign Affairs and Development and the Ministry of Defence each spend 3–4 per cent of the total, and the remaining ministries spend even less. The role of the Ministry of Finance in the determination of government R&D budgets is limited in the Netherlands. A large share of the budget involves commitments for years to come, such as financial obligations with respect to universities or the TNO. Policy discussions are usually confined to new policies. The Finance Ministry negotiates every year a total budget for every spending department. Within the confines of this total spending ministries have the autonomy to spend at will, as long as their plans do not conflict with general financial objectives of the Finance Ministry. Thus ministries should deliver on agreements with respect to general spending cuts. Also, they should not enter into spending commitments where future costs are unforeseeable due to the open-endedness of the arrangement. Financial windfalls or setbacks surfacing during the budget year and affecting a specific ministry are to be accommodated within the agreed budget of that ministry. The Finance Ministry does not scrutinize policies in detail *ex ante* on their financial consequences or their 'value for money', but it does keep an eye on the overall shifts in departmental budgets. For instance, the LNV claims to have realized important savings and efficiency gains through its reorganization of the agricultural knowledge infrastructure. The Ministry of Finance insists that these gains are made visible in the budget, either by a decrease in the total spend or by spending on new policies. The National Auditor's Office increasingly evaluates policies *ex post* regarding expediency, effectiveness and efficiency. Under the current political circumstances there is a considerable backing in Parliament for an increase in the budgets for technology policy and the Ministry of Finance supports this line through, for example, the development of tax instruments to support R&D in companies.

The Council of Ministers, comprising all 14 ministers in the Dutch government, has established a number of ministerial sub-councils to deal with specific policy fields, each consisting of ministers of the spending departments concerned. These sub-councils prepare legislation before it is submitted to the plenary meeting of the Council of Ministers. One of these sub-councils is the Ministerial Council for Science, Technology and Information Policy (RWTI). The activities of this Council are prepared at the official level within three Interdepartmental Assemblies, one each for science policy, for technology policy and for information policy (for which the Prime Minister holds responsibility). Policy documents are sent to Parliament for discussion. The Select Committees for Education and Sciences, for Science Policy and for Economic Affairs of both houses may then discuss policy with the minister involved. However, most discussion usually focuses on the general documents on science and technology policy which are part of the annual budget process. In general the role of Parliament in the development of R&D policy is limited.

Advisory Councils and Outside Players

A number of advisory councils are involved in the policy making process. Their task is not only to ensure the quality of policy decisions and to supply the policy agenda on S&T with new topics, but also to involve different interest groups in the policy making process and thereby to create support in society and among interest groups for policy decisions as they gradually evolve. Important councils are the Advisory Council for Science and Technology Policy (AWT) and the Sector Councils. The latter consist of representatives of relevant interest groups (for example, industries in their capacity as technology users), the science and technology communities and the government, and supply advice to ministers on specific issues of policy. Sector Councils currently in operation include the Councils for Health Research (RGO), for Agricultural Research (NRLO) and others. In addition the Royal Dutch Academy of Sciences (KNAW) provides advice on issues of basic research. Occasionally the Social Economic Council (SER), the Scientific Council for Government Policy (WRR) and the Central Planning Bureau (CPB) also contribute to the discussion through their studies and publications.

The main advisory committee in this field is the Advisory Council for Science and Technology Policy (AWT) which was established in 1990. The main responsibility of the AWT is to provide written advice to the ministers of OC&W and EZ, or via them to other ministers, on the science and technology policies to be pursued both domestically and in the international arena. The Council gives advice on request from ministers and on its own initiative covering topics relating to the generation, application and transfer of knowledge. Recent reports deal with issues such as the financing of higher education, the Netherlands as a location for innovative activity, technology policy and economic structure, the relationship between national and international S&T policies, the effects of various laws and policies on innovative activity, and the role and use of state-owned technological research institutes.

The Council acts independently of the scientific and technological communities and of the government. It takes account of such factors as the social context within which science is being practised and technology developed. The advisory reports of the AWT are as a rule made public. The Council consists of 9–12 members, all of whom are appointed by Royal Decree for a limited period of time. The members are drawn from various branches of society (academia, industry, and so on) and serve the Council in a personal capacity, not representing any vested interests. In general the recommendations of the AWT tend to have considerable influence on government policy. Most of its members have a high reputation. The present Chairman, for example, is the former research co-ordinator of Shell and former chairman of IRDAC (Industrial Research and Development Advisory Committee), advising the CEC on matters of R&D in industry. The position of the AWT within the system is relatively strong and its advice cannot be ignored. Ministers are obliged to submit the advice of the AWT, whether invited or not, to Parliament and to respond to it within three months.

Outside organizations with a general influence on technology policy are few. Most important is the Confederation of Netherlands Industry and Employers (VNO-NCW). All large Dutch companies are members of VNO-NCW, and so are three-quarters of companies with 100 to 500 employees, and the larger share of SMEs with fewer than 100 employees. About 160 branch organizations are also members. Currently communication between government circles and VNO-NCW is intensive, both formally and informally, the initiative coming from both sides. There are direct links to civil servants at EZ and OC&W, but also close connections to the Research Council (NWO), the Foundation for Technical Sciences (STW) and the AWT. It is quite remarkable that neither organizations of SMEs nor trade unions are very active in this field. Although there is currently a lot of support from policy makers for initiatives to stimulate innovation in SMEs, the interest and energy of SME organizations is concentrated on other topics. Trade unions claim to take an active interest in S&T policy but civil servants complain that they do not show much involvement in practice.

Figure 3.1 summarizes the Dutch science and technology structures. It reveals the main financial flows from the central government to the private sector, universities and research institutes. Of course, the reality is much

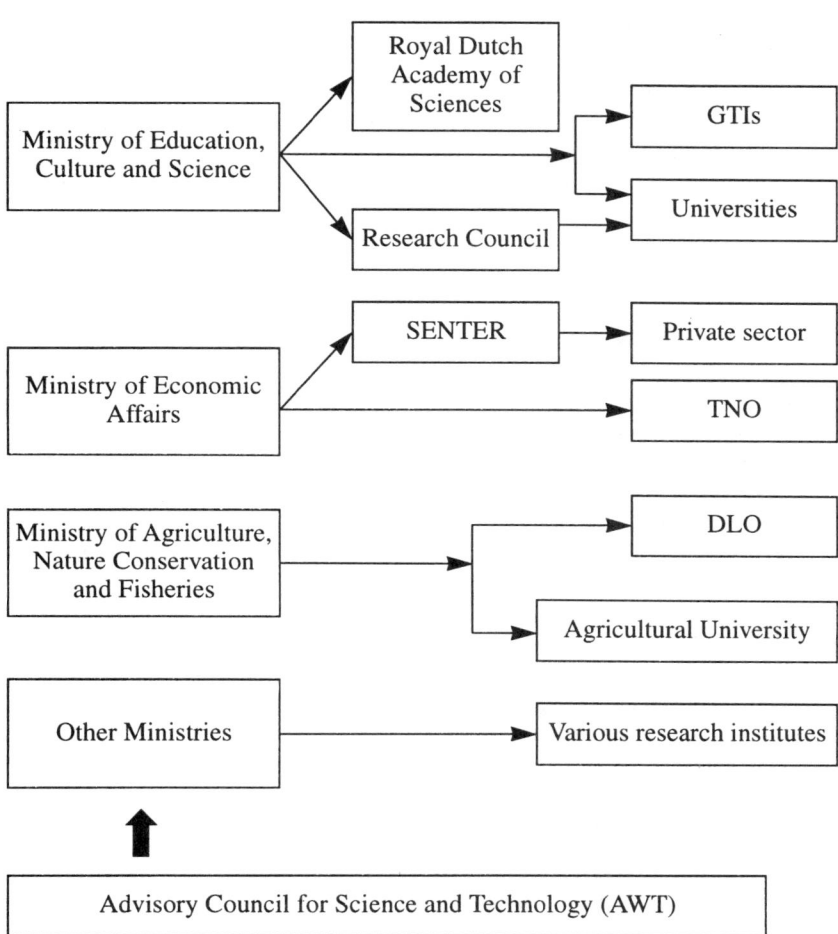

Figure 3.1   The structure of the Dutch public system of science and technology

more complex, and we recall that there exist all kinds of co-ordinating structures and cross-linkages that are perhaps typical of the Dutch 'poldermodel'.

Elements of the Debate: What is Going on in Dutch S&T?

Between 1987 and 1992 total expenditure on R&D in the Netherlands decreased from 2.28 to 1.98 per cent of GDP. This decrease was mainly due to a decline in private sector spending from 1.35 per cent to 0.96 per cent of GDP. Total R&D expenditure as a share of GDP in the Netherlands thereby came to lie well below the OECD average and even below the EU average. This has led to growing concern about the relative decline of the Dutch position in international comparisons and to an intensification of the discussion around S&T policy. A number of policies to stimulate business R&D through subsidies and tax rebates have come into effect in recent years.

A recovery of R&D investments has been recorded, especially in SMEs but not in multinational companies, and also an increase in business R&D financed from foreign sources. The current state of affairs with regard to Dutch R&D can be summarized as follows:

1. The public knowledge infrastructure, the system of universities and public sector research institutions, performs fairly well. The quality is in most disciplines above average. Public investment in the knowledge infrastructure as a whole is comparatively high, but happens to be concentrated in fields of knowledge that are not directly relevant to technological performance. There are fears that there is too strong a bias towards languages and social sciences and a lack of emphasis on natural sciences and engineering. A substantial shortage of qualified engineers is projected for the near future.

2. The share of government budgets allocated to support business R&D are low by comparison, although they increased between 1991 and 1995 (one of the reasons for their being low is the limited defence R&D spending). Government support for R&D within the public knowledge infrastructure is decreasing in the Netherlands, whereas it is increasing in most competitor nations.

3. Most concern surrounds business R&D in the Netherlands. Though there has been a recovery in recent years, the Netherlands has lost its place as a front runner in international league tables. Whereas the public sector in the Netherlands spends just over 1 per cent of GDP on R&D, which is more than in Germany, Japan, the UK and the US, the private sector also spends just over 1 per cent, far less than is spent in the four countries mentioned. Dutch R&D intensity is lower than average in practically all manufacturing industries.

4. Regarding the output of technology, it appears that most Dutch patenting activity is concentrated in the low and medium technology industries.

5. There are fears that the Netherlands has obtained less advantage from new technologies as a source of economic growth and productivity than competitor countries. In the Netherlands, unlike abroad, high-tech sectors such as computers and aerospace have failed to contribute substantially to employment growth.

6. R&D in Dutch business enterprises is focused more on research and less on development. Dutch companies seem to succeed less than foreign competitors in succesful product launches or improved production processes. Also, Dutch companies on average generate less of their turnover from new products and processes (EZ, 1997).

Although not all these points are recognized by all parties involved or judged important to the same degree by everyone, there is a consensus that the Netherlands could do better and that government involvement is called for. There is some discussion regarding the extent to which current trends are related to structural characteristics and trends. Three factors are deemed important in this context.

First, the Dutch economy is fairly small and has a relative specialization in industries such as agribusiness, food processing, chemical processing, distribution and services. Dutch manufacturing is concentrated in comparatively medium- and low-tech industries for which world markets are growing fairly slowly and prices are more or less stagnant. The Netherlands is underrepresented in technology-producing industries where new technology is embodied in the final product (for example, semiconductors, medical instruments, pharmaceuticals). It is overrepresented in technology – using industries that are dependent upon others as suppliers of process innovations. This explains why Dutch firms are large technology importers, paying about 0.5 per cent of GDP in licence fees to foreign sources, whereas they export technology only to a value of around 0.2 per cent of GDP.

Second, the Dutch economy is heavily dependent for its business R&D upon a small number of large multinationals. Five transnational corporations (TNCs) – Philips, Shell, Unilever, Akzo and DSM – accounted for between 55 per cent and 60 per cent of Dutch business R&D expenditures in 1993 (down from 71 per cent in 1970). A marginal decrease in the R&D expenditure of one of those five has a large impact on the total of Dutch R&D activity. The R&D intensity of these five TNCs has dropped by almost one-half, from 0.9 per cent of GDP in 1987 to 0.5 per cent in 1993. A total of 55 companies (of which 17 are foreign multinationals) spends around 94 per cent of total business R&D costs. There are 240 firms which spend more than 1 million guilders per year on R&D in the Netherlands (Minne, 1995). Tax rebates stimulating R&D investments in particular have caught on among companies with fewer than 1000 employees and among high-tech start ups, and thereby have reduced dependence on multinationals.

Third, multinationals around the world are increasingly becoming footloose as they move more and more functions, including research and product development, closer to the markets they serve. Dutch multinationals are gradually transferring R&D facilities abroad, especially for near market R&D,

but this has not been fully compensated by foreign TNCs moving in. The share of R&D that is financed from foreign sources, for example TNCs which locate their research facilities in the Netherlands, has decreased sharply since the mid-1980s, though there is some recovery in foreign-financed business R&D in recent years. It is argued that the costs of business R&D in the Netherlands are high relative to quality, and that locally available technological potential falls short of demand.

Nevertheless, these structural traits and developments cannot fully explain the Dutch decline. There is evidence to suggest that Dutch firms on average invest too little in R&D, are slow adopters of advanced production technologies, are sluggish in tailoring products to markets, in adopting managerial and organizational innovations (for example, low integration of production and other functions, leading to longer 'time to market'), and put less effort in to vertical co-operation. There is a constant debate going on about the truth of these claims as well as the likely causes: lack of incentives, policy failures, skills shortages, labour market rigidities, for example.

3.3   PHILOSOPHIES

The Views of the Ministry of Education, Culture and Science

The Ministry of Education, Culture and Science (OC&W) is in charge of the care of the Dutch scientific infrastructure and the maintenance of the national knowledge and skills base. The starting point of its philosophy is the recognition that there is no alternative to investment in knowledge if the Netherlands wants to fulfil its ambitions with respect to competitiveness, standard of living and wellbeing. Among the main global developments that the Netherlands has to deal with, it cites:

- increasing competition from Asian countries where there is strong growth of scientific and technical capacity;
- the increasing importance of access to scientific and technical knowledge, and the ability to exploit knowledge to generate innovations which lead to increases in value added;
- the growing importance of factors other than R&D for successful innovation, such as education, social, ecological and cultural factors, marketing and management;
- the transformation of production processes themselves, as well as the development and diffusion of knowledge itself, as a consequence of the development of information and communication technologies.

In addition there are issues specific to circumstances in the Netherlands as a small and densely populated country which demand an extra investment in the development of knowledge, such as environmental pressures, scarcity of space, the openness of the economy and social cohesion in a multicultural society (OC&W, 1996).

Given that the aim is to create a knowledge-intensive society, there are three issues which are the central drivers of the policy making process at OC&W: (a) the necessity to make choices; (b) the problem of getting commitment from society; (c) the problem of reaching the top level in science. We shall deal with each of these issues in turn.

Making choices

Because the Netherlands is a small country, has limited funds available and is part of a globalizing and integrating world, it is necessary to make choices, that is, to promote and support some areas of scientific advance at the expense of others. A basic provision of every type of scientific and technical knowledge should be available in the Netherlands, if only to be able to absorb and use knowledge developed elsewhere in the world. For investment other than that, priorities have to be determined, not only within but also between fields of science. The conclusion that choices have to be made was reached more than a decade ago. Since then a slow process of acceptance of the consequences of this recognition has been set in motion. In 1992, OC&W installed the Consultative Committee on Government Surveys (OCV), its brief being to identify both scientific areas of priority and of 'posteriority' (that is, those to which flows of public resources would gradually be curtailed). The OCV distinguishes between three reasons why research may be rated of vital interest to society: (a) because it is essential to maintain high standards in academic education; (b) because it is important for the functioning of both private and public (profit and non-profit) sectors of society; (c) because of its importance for the solution of broad problems in society and improvement of the quality of life in the future. This approach expresses a concern that is not limited to the competitive position of the Netherlands in a global economy. Explicit attention is given to the role that scientific development can have in solving problems of economic vitality, social cohesion, ecological diversity and sustainability. Areas of interest from a societal point of view are, for instance, health problems related to the ageing of the population; containment of criminal behaviour; maintenance of physical infrastructure; and the mobility and vitality of urban areas.

Important in the OCV surveys is the involvement of both suppliers of new knowledge and representatives of the user communities (private as well as public sector). It is felt that the process of surveying, as it increases awareness of the issues at stake among a large variety of players involved, is at least as important as the outcomes. The survey process is a way to develop strategies at the national level that are supported by all parts of the community. A main instrument in the OCV surveys is scenario analysis. Scenarios are used in workshops in which a broad range of stakeholders participate. Here S&T strategies are developed and evaluated against the background of various possible futures. Scenarios are used to bring to the fore fundamental uncertainties about future developments. They foster strategic thinking and help to evaluate the benefits, costs and risks of different courses of action and their robustness in the face of critical uncertainties. The Dutch OCV surveys differ from most forward-looking exercises in other countries in their broadness of scope. Contrary to most national survey activities, the OCV does not focus so much on identification of future key technologies, for example, by using Delphi-methods; the fields under consideration include also basic science and social and cultural disciplines.

The OCV delivered its final report in 1996. It appeared particularly difficult to indicate areas for 'posteriorization' and to recommend choices between, rather than within, disciplines. Nevertheless, a long list of recommendations has resulted, some on the development of knowledge within 12 different themes, aimed at dealing with practical problems and generating innovations, others on the further development within a number of scientific disciplines (natural and technical sciences, information technology, social and behavioural sciences, law and humanities). After the publication of the OCV's final report and after consultations with other ministries, with organizations in the science base (universities, NWO, KNAW, TNO and so on), and with Sector Councils, the conclusions have found their way into the science budget. OC&W has called upon the science base to shift its priorities in line with the recommendations of the OCV within existing budget outlays. For some themes a limited extra budget has been made available. The OCV has been dismantled, but the survey process will be carried on by the AWT.

KNOWLEDGE THEMES

The OCV signals four main drivers that determine the future needs for knowledge and expertise in the Netherlands: (a) information and communication in the knowledge society; (b) the aspiration for a competitive and sustainable economy; (c) internationalization and regionalization; (d) improvement of the quality of life. On the basis of an analysis of these four driving forces, the OCV has identified 12 themes of knowledge which are important to tackle the issues that the Netherlands will face in the near future and has recommended directions for investments on every theme:

1. *Information and communication infrastructure* Invest in a long-term research programme to develop innovations for the 'electronic highway' and in research aimed at evaluating the consequences for society of further ICT development.
2. *Human capital* Invest in research into education needs and improvement of learning processes in the knowledge-intensive society.
3. *Agrifood industry* Reinforce research aimed at renewal and quality improvement in agriculture and the food sector.
4. *Services* Invest in research for productivity growth and renewal in services.
5. *National research initiative 'Factor 4'* Invest in innovations aimed at doubling prosperity whilst reducing environmental pressure by one-half.
6. *Integrated use of space* Reinforce research into optimization of logistics and into spatial planning and public administration.
7. *New economic activities and innovation* Develop research for knowledge-intensive production in a globalizing world.
8. *Internationalization and regionalization* Invest in research on international law and on cultures of regions that are of special interest to the Netherlands, and into research into Dutch language and culture.
9. *Social cohesion* Invest in research into opportunities for and threats to social cohesion in Dutch society.
10. *Health* Reinforce research into the quality of life and the effectiveness of health care.
11. *Global issues* Aim at a targeted, visible contribution to knowledge of global environmental problems (climate, coastal zones, biodiversity).
12. *Energy* Invest in long-term research into sustainable, renewable energy resources and in medium-term research aimed at energy savings and exploitation of Dutch expertise in the area of natural gas.

Implementation of the agreed policies and budgets depends on the co-operation of organizations in the science base. In intensive discussions between the Minister of OC&W and representatives of those organizations, a combination of persuasion and coercion is used to ensure compliance. The universities are organized in the Association of Netherlands Universities (VSNU). With respect to the recommendations of the OCV, for instance, the

Minister has agreed with the VSNU that every university will report to him at a certain moment in time on which recommendation has been implemented in what way, who is responsible and how much manpower is involved. If no agreement with the VSNU can be reached the Minister has the authority to allocate 3 per cent of the budget. Against the NWO, the KNAW and the TNO he is in a stronger position: he can reject the entire budget. *Vis-à-vis* other departments (which finance parts of the science budget) he has no powers to enforce compliance if they intend to deviate from the agreed budget.

Getting the private sector and society at large more involved

A second issue that has been on the minds of policy makers at OC&W for several years is the relationship between science and society. There is the paradoxical situation that the growing importance of knowledge for the smooth functioning of society is widely recognized but that budgets available for investments in science and technology are under constant (political) pressure. At the root of this paradox is the mismatch between the supply of and the demand for knowledge. Investments in knowledge are perceived to have a low return because the science base is too little demand orientated and the user community does not articulate its needs clearly enough. This has brought up the question of what can and should be the influence of society on the research agenda in the science base. The OC&W takes the line that society should be made to feel itself 'co-owner' of public research. Co-ownership means that groups within society, whether sectors of industry, charities or other private sector organizations, or the public at large should feel involved in the research process. This implies various things. First, potential stakeholders should be given an opportunity to contribute to the determination of research priorities, at least at the aggregate level. This happens, for instance, through their involvement in foresight activities. Second, groups in society should be helped to take advantage of research results. This can be accomplished by improving communication between researchers and potential users and by fostering long-term relationships between research groups and stakeholders in society. Third, co-ownership also, in the view of the OC&W, implies a financial commitment. If research is to benefit certain groups in society, they should take the responsibility to co-finance the research programmes. The Ministry then sees it as its role to facilitate the bringing together of these groups and to build an institutional structure in which co-operation can materialize. Examples are the Leading Technological Institutes' (TTIs) initiative and certain research programmes, such as Economy, Ecology and Technology (EET). Not surprisingly, many organizations in Dutch society, for instance the employers' organization, are quite positive about the first two points, but are less enthusiastic about the third.

WHO PAYS FOR AND WHO GOVERNS DUTCH UNIVERSITIES?

Basic R&D is performed mainly in universities. Besides basic research, externally financed research activities of a strategic or applied nature are becoming more and more important sources of finance for universities. In 1995, Dutch universities received 57 per cent of their funds through OC&W as a lump sum linked to student numbers (the so called primary stream of funds), 17 per cent through the Research Council (NWO) and 26 per cent through contract research (OC&W, 1997b). The primary stream of funds is meant to cover the costs of education and basic research. Allocation of these funds to various activities is essentially up to the universities themselves.

The Research Council finances mainly basic and strategic research. Its budget is divided among programmes and projects in universities and maintenance of educational networks for PhD students. To allocate these funds NWO organizes tenders and applies quality criteria to judge research proposals and evaluates the utility of prospective results. In 1995, only 20 per cent of contract research in universities was financed by business enterprises; 30 per cent was financed by government, 33 per cent by non-profit organizations and 17 per cent from abroad (OC&W, 1997b).

When compared internationally, Dutch universities receive a large share of their funds as a lump sum (0.31 per cent of GDP in 1994, as against 0.14 per cent in the UK, for instance). The funds that flow through the Research Council are comparatively small; this type of finance makes up 25–30 per cent of university funding in the UK, France, Belgium and Denmark. Contract research in Dutch universities is in line with other European countries, but much smaller than in the US.

Because Dutch universities receive a large share of their finances as a lump sum per student, they are relatively autonomous in the determination of their research priorities. Moreover, fields of research are determined more by the popularity of a discipline among students than by the needs of society at large. The OC&W feels that this situation is more and more undesirable. First, budgets are constantly tightened and pressures to spend tax payers' money to the benefit of tax payers' interests are mounting. Second, one of the major worries of the government is the apparent mismatch of demand and supply of Dutch research and the poor connections between the private and the public R&D networks. Third, an extensive foresight process was concluded in 1996 and has resulted in numerous recommendations, and it is felt that these recommendations should have an effect on publicly funded research programmes.

The Ministry is therefore looking for ways to get a better grip on the universities and on their research programmes. The objective is to get university research more orientated towards societal needs and to increase the involvement of (parts of) society and business in university research. For instance, in exchange for co-responsibility and financial involvement, the private sector is offered influence on research programmes. The mechanisms through which this will be effected are still being developed. One initiative is the creation of Leading Technological Institutes (TTIs), four of which were created in 1997 (in the fields of nutrition sciences, metal technologies, polymers and telematics). Knowledge institutions and businesses each provide 25 per cent of the finance for TTIs, and government supplies the remaining 50 per cent. Another initiative is the establishments of 'research schools' within the universities. Research schools are meant to create a critical mass of high-level academic research in a certain field within one university. They have to go through an arduous procedure for recognition by the KNAW, but once they are recognized they can tender for additional funding. This additional funding is obtained by reducing or earmarking the primary stream of funds of the universities. Further initiatives include a substantive transfer of funds from the primary stream to the Research Council, and the obligation on universities to develop plans for the implementation of the recommendations that have come out of the foresight exercise.

In the discussion about how to guide the various streams of funds that finance the universities, the Advisory Council for Science and Technology Policy (AWT) argues for a clear distinction between different streams of funds and for 'one stream of funds for one objective'. The primary task of universities is to educate people. Research is a necessary activity that enables universities to provide first-class education to students. This sort of research should be financed through the primary stream of funds and the spending of these monies should be at the discretion of the universities themselves. For activities financed through primary funding, a system of quality control should be agreed with the government to evaluate the results of the university as an educational establishment. Research Council funds should be aimed at stimulating the very best research, where the criterion is purely academic excellence. Government should refrain from attaching conditions to the content of research. Research for which there is a commercial market should be directly financed through contracts. Research to contribute to the resolution of long-term societal needs (for which there is no market) should be paid for by various ministries, where the Research Council can contribute to the efficient allocation of the funds.

Top-quality research

The third item on the agenda of OC&W is the fostering of top-quality research. The productivity of the Dutch public knowledge infrastructure is good: the Netherlands figures in tenth place in the world's league table of productivity in academic research output and seventh in terms of impact (measured by numbers of citations). The quality of applied research as performed by TNO and the Large Technological Institutes is monitored by means of regular evaluations by international commissions. However, it is felt that the average is overrepresented and that genuine excellence merits more stimulation. Means to do so are found: in the establishment of research schools in universities to create a critical mass of excellent researchers and in providing the best ones among them with extra funds; in strengthening and extending the role of the Research Council; in the foundation of Leading Technological Institutes (TTI) for top-level applied research; and in a well-funded system of fellowships aimed at providing very promising researchers and top scientists with the necessary means to develop their careers and their lines of research.

The Philosophies of the Ministry of Economic Affairs

The Ministry of Economic Affairs (EZ) takes care of the interests of Dutch industry (except agriculture). During the 1970s EZ conducted an active industrial policy and on several occasions provided extensive support to Dutch manufacturing firms in trouble. This defensive approach to policy, directed predominantly at weaker elements of manufacturing industry, was discredited and abandoned when the Dutch shipbuilder RSV went bankrupt despite millions of taxpayer's money. This induced a shift in the mission and policy orientation of EZ and from that period onward a less reactive and more proactive approach has been followed. Technology policy as we currently know it started to be developed at the beginning of the 1980s and came to play a key role among EZ's policies. The objective of EZ R&D policy is to promote the competitiveness of Dutch industry by providing access to sources of technology and by providing support to technology development. At first technology policy was focused upon development and innovation in a limited number of technologies and product markets which were deemed important for the revival of Dutch industry after the crisis of the 1970s. Gradually the scope was broadened to include technology diffusion, social acceptance of technologies and the role of technology in tackling problems in society at large. More attention was paid to generic conditions and internationalization became an important theme. Recently more attention has been paid to the relationship between the public knowledge infrastructure and private industry, to co-operation in R&D, and to problems of matching demand for and supply of skills on the labour market.

On the one hand, EZ acts as a facilitator in the communication among business enterprises and between industry and the state; and on the other hand, it allocates financial support to enterprises through various channels. As a facilitator EZ stimulates and supports the establishment of networks and the formation of enterprise clusters. The latter are coherent groups usually comprising one larger technologically advanced firm, a range of suppliers and parts of the public knowledge infrastructure. Furthermore, EZ tries to optimize the match of supply of and demand for knowledge and technology. It therefore stimulates co-operation between firms and public sector technology providers. Allocation of support by EZ is relatively small scale. Providing large-scale financial support not only falls under the restrictions of EU regulations but is also politically not viable. Given the importance of bringing down the public sector borrowing requirement (in view of the requirements for joining the EMU, as laid down in the Maastricht Treaty, and for staying in, as laid down in the 'stability pact') and the ferocity of the yearly debates around the recurring rounds of budget cuts, large-scale support would be politically sensitive. Moreover, under the favourable economic circumstances of the last couple of years the idea of tax cuts has been rising on the political agenda again. However, though support is not handed out as liberally as before the fall of RSV, the pendulum is swinging back again. Arguments that there is a national interest in maintaining high-tech manufacturing industries, and that even massive support is justified if it averts the risk of losing a valuable asset which generates important externalities and which can never be regained, are gaining increasing support. The aircraft manufacturer Fokker only went out of business after a number of rescue operations and several rounds of financial injections from the state's purse had failed. Philips receives, year in, year out, very substantial amounts of support (around 1 hundred million guilders) for its R&D activities in electronics.

A consequence of European integration and a shift of focus to the international scene has been a heightened awareness of 'policy competition' between countries and the increasing weight of tit-for-tat arguments as a legitimation of policy initiatives. In general it is felt to be a task of EZ to try to influence European policy to the benefit of Dutch companies and to resist (specific) support policies for companies elsewhere in Europe. However, if unfair competition results, EZ should introduce compensating measures. Often the 'level playing field' is the only credible argument that can be raised to match foreign policies.

TECHNOLOGY RADAR

In parallel to the survey activities of the OCV there have been other Technology Foresight exercises in recent years investigating the medium-term needs of the Dutch business community in particular. One of these, the Technology Radar, was commissioned by the Ministry of Economic Affairs and aimed, first, to identify technology fields that are likely to be of strategic importance to Dutch business and industry within the next ten years, and second, to investigate whether sufficient knowledge buildup is taking place in these fields of strategic importance.

The Technology Radar identifies 15 key technologies and, for every one of those, maps out characteristics and trends, demand and supply, and issues that affect the use of these technologies in business. The technology and innovation system which propels progress in every one of these technologies has its specific problems, but there are a number of recurring features worth mentioning:

- The interaction of suppliers and demanders of technology in the Netherlands needs improvement. Better and more frequent communication between research institutes, ministries, industry associations and individual companies is called for.
- Graduates in technical fields from Dutch universities are well trained and of a high quality. However, industry signals

two related weaknesses. First, much of the work in companies is carried out by teams of people from different
disciplinary backgrounds. Graduates usually lack experience of work in multidisciplinary teams. Second, industry
wants graduates to be more broadly educated. Dutch graduates are quite knowledgeable in their own field but have
had limited exposure to other disciplines.

- There is a decline in interest among young people in technical studies and careers. There are concerns about the supply
of engineers, especially computer science graduates, information and communication technologists, electrotechnical
engineers, graduates in process and materials technologies, and some newer areas of pharmacology.

- Industry sees university research as fragmented. Relevant
knowledge is scattered throughout departments and universities. Research schools partly remedy this problem but
as they are monodisciplinary in nature they do not
accomodate industry's multidisciplinary research needs.

- Universities are not organized in a way that permits them
to react quickly to the rapidly changing research needs of
companies engaged in the development of products with
short life cycles. Their primary focus is on monodisciplinary,
long-term, basic research. Improvements in responsiveness could be obtained by changes in incentives to
researchers, for example by encouraging researchers to
obtain private sector funding.

- Companies in an industry are mostly unable to come up
with a single consistent view of their research priorities for
a particular technology field.

The Views of the Other Players

The Ministry of Agriculture (LNV) pursues an R&D policy in the interest of
the agricultural sector, nature conservation and rural development, and fisheries. Currently R&D policy in agriculture is in a state of change. For years the
emphasis in agricultural research and development has been on generating
productivity improvements in the production of bulk commodities. This was
efficiently organized through a system that identified and produced new
techniques centrally and diffused them top-down. New market conditions,
international competition and the need to develop more sustainable methods
of production forced a change in technological development toward more

quality-orientated production and diversity in development. Demand for variety, a focus on durability and global competition require a new organization of the knowledge system with more regard for bottom-up initiatives, a more active role for primary producers, improved communication between sources and users of technology, a more pronounced international orientation and a change from input to contract financing of research.

CHANGING PHILOSOPHIES: THE CASE OF INNOVATION SUBSIDIES FOR AGRICULTURE

A prime example of corporatism in the Dutch economy used to be the functioning of the agricultural sector. As in most countries, the sector consists mainly of a large number of family businesses producing a homogeneous output. To counter the economic hardships of having to function under full competition, the sector has always managed to organize its collective interests very well, for instance through sales co-operatives, sector and commodity boards, and various sector organizations. Also, farming interests have always been very well represented in political spheres. Until 1994 the Minister of Agriculture used to be a representative of the sector. A substantial part of technological development in agriculture is financed directly by the Ministry of Agriculture, part is financed collectively through levies paid by farmers to commodity boards.

Currently conditions for the agricultural sector are changing. Segmentation of markets offers producers the opportunity to specialize and to differentiate their product from that of competitors. These developments induce an increase in competition within the sector. As interests diverge, collective organizations have come under pressure (the Agricultural Industry Board has been dissolved) and the political impact of the 'green front' is fading. Opportunities for differentiation stimulate entrepreneurship and innovation on farms, but also induce within-sector competition and increasingly prohibit the sharing of new technology and technical knowledge. At the same time ideas on the role of government are changing. Where in the past the Ministry of Agriculture saw it as its task actively to promote the collective interests of Dutch agriculture, it now increasingly sees itself in a more enabling role and is shifting responsibilities for business development to the private sector.

These processes of change are similar to current changes in other sectors and ministries. They are reflected in new policies. In the 1980s the idea that technological change in agriculture should be stimulated on the demand side caught on and policy instruments to subsidize innovative projects on farms were introduced. In 1997, a regulation from the beginning of the 1990s under which subsidies were attributed was replaced by a new regulation. The differences between the old and the new policy illustrate the changing philosophies:

1. The aim of the old regulation was to 'stimulate innovative activities' as a goal in itself; the aim of the new regulation is to 'increase the competitive and market power'.
2. The old policy subsidized agricultural enterprises only, the new policy subsidizes all firms operating within agro-industrial chains. This reflects the increased emphasis on, and increased expectations from, co-operation between vertically related enterprises, as well as more attention paid to opportunities in the field of organizational innovation and innovation in comprehensive production systems.
3. The old regulation supported all projects that fulfilled certain criteria; the new regulation works with a tender system where projects compete against each other.
4. Previously, projects were evaluated by a committee of civil servants and representatives of sector organizations, whereas now project evaluation is in the hands of an independent commission. Civil servants may be inclined to judge projects with an eye to the current policy agenda and industry representatives may want to maximize sector income. An independent commission is supposed to balance tax payer's versus sectoral interests better and have more regard for value for money for society as a whole.
5. Under the old regulation there was only a subsidy; under the new regulation there is also assistance in the development of the project plan. It is increasingly recognized that the needs of entrepreneurs in small businesses are not only, or not even predominantly, of a financial nature but also have to do with lack of time, information and organizational skills.

The Advisory Council for Science and Technology Policy (AWT) stresses that government technology policy should seek alliance with initiatives which originate in business. Government should concentrate on providing favourable conditions and infrastructure for R&D and should thereby provide positive incentives to business. Rather than be involved in innovation itself, government should assist firms in 'technology asset management', finding and using the right ingredients for technological renewal, no matter whether acquired internally or externally. The AWT also underlines the fact that the Dutch economy is extremely open and that it is therefore of prime importance to ensure that conditions in the Netherlands will be attractive to firms. Attractiveness requires sound economic policies bolstered by prudent R&D policy. The AWT distinguishes between three types of companies: TNCs, large domestic exporters and SMEs. TNCs have access to international sources of technology and sufficient financial resources. For them high-quality education at all levels and local availability of engineering skills is of foremost importance. Attracting them may also require matching foreign policies. Large domestic exporters are more dependent on the national knowledge infrastructure and national sources of finance. SMEs in addition need government support in particular to gain access to the sources of technological knowledge that are available, to make use of technologies that have the potential to be profitable if they can be readily identified and their implementation can be organized. The technological development of part of the SME sector is a cause for concern. The AWT signals the changes in the relationships between large firms and their often smaller suppliers. Large domestic and foreign firms are ever more reorganized into independent business units, subcontracting out parts of the production process and withdrawing to key activities. This makes access to highly sophisticated supplier firms an important condition for location. There are indications that parts of the Dutch SME sector has trouble reaching the required sophistication, that is, transforming from jobber to co-maker.

SUMMARY OF RECENT INFLUENTIAL S&T POLICY RECOMMENDATIONS

Advisory Council for Science and Technology Policy (AWT)

1. Technology policy should take account of the different needs of multinational companies, large national firms and SMEs (see main text).
2. Public policy should link up to initiatives in the private sector and refrain from selecting content; policy should have a generic character. Therefore technology policy instruments that support only innovative initiatives within certain fields chosen by government should be abolished.

3. Innovation in SMEs needs extra attention. SMEs have important needs with regard to the management of technological assets; they should be supported in finding necessary information and making optimal use of available resources. This should be done by improving access to innovation development credit (IOP) and by extending the services of innovation centres.
4. The match between knowledge infrastructure and private sector knowledge needs must be improved. Knowledge institutes should develop research strategies and be judged and funded on the basis of those. Public funding of research institutes can be tied to their success on the market. Public contributions to covering the costs of strategic research contracts that firms grant to public research institutes can stimulate a better match.
5. Vocational training at all levels should be improved. Students should be prepared for technological asset management: learning in networks. In engineering courses more attention to design is necessary.
6. Technology should be a consideration in all fields of policy making. The possibly negative impact of (new) legislation on innovation in the private sector deserves constant attention. Government can foster innovation through its investment expenditures. Government can co-ordinate the introduction and exploitation of network technologies in which large numbers of stakeholders are involved.

Social Economic Council (SER)

1. Public research institutions should contribute more to meeting the technology needs of the private sector. To encourage this, public funding of research institutes could be made to depend on successes in attracting contract research.
2. Further budget cuts affecting education, except in the case of efficiency improvements, are not advisable. Nevertheless, funding could be made more conditional upon institutions reacting to labour market needs.
3. Cooperation between higher education and private businesses must be improved. Student enrolment in sciences and engineering must be stimulated.
4. Allocative efficiency of product and labour markets should be improved.

5. R&D subsidies to firms should be scrutinized with respect to their effectiveness. Results should be compared to returns that could be obtained through alternative investments, for example, in research infrastructure or education.
6. In exceptional cases, when there is an overriding strategic interest for the Dutch economy, the public sector should take an initiative to actively support clustering and technological partnership in the private sector.
7. SME's should be stimulated to engage in strategic partnering if, due to lack of scale economies, own development is unprofitable or too risky.

Sources: AWT (1994), SER (1995)

The Confederation of Netherlands Industry and Employers (VNO-NCW) stresses foremost that in matters of technological development industry should largely determine how finance is spent. Government subsidies to public knowledge institutions and research laboratories could be better linked to the needs of industry, thereby strengthening the link between S&T and wealth creation. VNO-NCW lobbies for more support and incentives but for less intervention. Broad generic support policies are needed but competition in the market should not be interfered with and protectionism is likely to hurt trade interests in the long term. VNO-NCW has two priorities regarding government policy at the moment. First, government should stimulate innovation in enterprises on a continuous basis through the use of 'generic' instruments. Access to these instruments should be as broad as possible; government should not apply restrictions, for example by putting a size limit to potential applicants or by establishing that only projects in designated fields are eligible for support. Also, government should match foreign policies. If foreign governments give financial support to R&D in firms, so should the Dutch government. Second, the public knowledge infrastructure should be reorganized to improve the return on investment in terms of technologies that have the potential to be used commercially. This implies a concentration of funds in fields of promise or excellence. Also, universities should be stimulated to work more for industry (for example, through contract research). Moreover, knowledge that is available should be diffused more widely and effectively. This suggests improved communication between producers and users of knowledge.

HOW TO DEAL WITH HYBRID ORGANIZATIONS: TNO, DLO AND THE GTIs.

The Netherlands Organization for Applied Scientific Research (TNO) comprises 15 research institutes, employing around 4400 people. It specializes in applied (mostly technical) research to the benefit of Dutch society and industry and is comparable to the German Fraunhofer Institutes. It participates in all four Leading Technological Institutes that are now in the process of being established. In 1995 the research budget of TNO was 561 million guilders, of which 30 per cent came from private business and 14 per cent from abroad. The rest of the funds came from government, either as a subsidy or through research contracts. TNO receives a basic subsidy from OC&W and earmarked (targeted) subsidies from OC&W and other ministries (CBS, 1997).

The Agricultural Research Department (DLO) employs about 3000 people and had a turnover of around 440 million guilders in 1998. More than 75 per cent of its income comes from the Ministry of Agriculture (LNV). At the moment DLO is being given legal status independent of the state, a road that TNO followed before. In the future, DLO will come under a holding together with the agricultural university and the experimental stations that cover the whole of the agro-industrial knowledge trajectory. In addition to TNO and DLO, there are five GTIs, large technological institutes, doing applied research for the public and private sector. They are active in the areas of energy research, geotechnics, maritime research and engineering, aerospace and hydrodynamics. Together they had a turnover of 400 million guilders in 1996 and employed around 2500 people. Government subsidies account for between 10 per cent and 65 per cent of their income. Public sector contracts are important sources of income for most of them.

Over the past decade or so, all these research institutes have been gradually pushed from the protective bosom of government in the direction of the jungle of the free market. Institutes developed into 'hybrids': partly state dependent, partly market driven. Subsidies (input finance) are step by step being replaced by contracts (output finance). This has had a number of well-known positive effects: more efficiency, more accountability and lower costs for the public purse. The imposed discipline of the market is supposed also to have benefits for the execution of the public tasks of these institutes. However, as the process progresses, the disadvantages of these developments are becoming ever more apparent. As the freedom of the institutes to compete in the market was increased while government budgets decreased at the same time, public sector research institutes started to compete with their private sector counterparts. Complaints of unfair competition were voiced: public sector research institutes receive tax advantages, can use their subsidies to co-finance their market activities, sometimes have access to cheaper capital through the government, and have often obtained their assets at a discount in the process of privatization. In addition, the more that public research institutes indulge in commercial activities, the more they are inclined to protect their knowledge from the public eye, even when this knowledge has been developed with state funding.

These complaints and problems have spurred a discussion in the Netherlands on the conditions under which these hybrid organizations should operate (see, for example, Werkgroep Markt en Overheid, 1997, and AWT, 1996). Hybrid organizations combine elements of a mission-driven (task) organization and a market-driven organization. A task organization should within the field of its mission develop new knowledge and technology, as much as possible with the needs of potential users in mind. A market organization should first and foremost apply its existing knowledge to solve the problems of its customers; it should also have some leeway to invest in the development of strategic expertise to be prepared for the demands of the market of the future (technology readiness). Largely task organizations and market organizations should be separated legally, organizationally and financially. On the one hand, task organizations should not go commercial but live off subsidies, which should be sufficient to fulfil the mission of the institute. They should concentrate on technology development, in co-operation with private sector firms and market driven research organizations, possibly even under the umbrella of a university. Market organizations, on the other hand, should act as knowledge transfer organizations, getting as much finance as possible out of private and public sector contracts.

There are two main arguments for separating task and market organizations. The first revolves around unfair competition, as mentioned above. The second has to do with culture: a market driven organization functions with an enterprise culture, has other criteria for success, and appeals to other capacities in employees than a task- or mission-driven organization. Impressions so far suggest that attempts to combine a commercially orientated culture with an academic, mission-orientated culture in one organization tend to lead to suboptimal performance, both in the market and in task fulfilment. Too strict a separation, however, is not desirable either, as it limits the transfer of new knowledge through the system and hampers communication between mission-orientated researchers and potential user communities. Government should maintain task organizations active in fields that are of direct relevance to the state, such as safety and security issues, environmental care, infrastructure and defence. In research, government should maintain facilities at the basic end of the technology spectrum. In addition, however, government has a responsibility in financing part of market-driven research organizations. It is recognized that it is not possible for publicly owned market-driven research institutes to raise funds in the market for the development of strategic expertise aimed at satisfying future demand. The time horizon of the market is too short and the public good character of strategic knowledge makes the private sector reluctant to invest. Thus government has a role in subsidizing 'technology readiness' in these type of market organizations in the interests of Dutch business. To stimulate market orientation, however, a major part of the funds should be made available through subsidies to the demand side, that is, to firms that spend these funds on research contracts, for which research institutes compete.

The discussion about the appropriate relationship between government and market in the research sector has not yet been concluded and its translation into policies has just started. At present, TNO, DLO and some of the GTIs have mission statements that cover the whole spectrum from basic technology development to near market research and advisory services. They compensate their losses on market activities with their subsidies and they attempt to combine a mission-driven and a commercial culture. It is likely that in the near future some of these research conglomerates will be restructured on the basis of the arguments advanced in this ongoing discussion.

3.4 POLICIES

The main policy concern of the Ministry of Education, Culture and Science is the strengthening of the science base by reallocating funds among recipients in the knowledge infrastructure. Given the difficulties of setting priorities and the position of vested interests, this is a long-term project. An important instrument is the implementation of changes in the administrative structure of knowledge institutes (universities, TNO, GTIs, and so on). In dealing with TNO and the GTIs, finance is increasingly tied to projects and provided on a contract basis. In dealing with universities there is an attempt gradually to detach funds for research purposes from student numbers and to dissociate decision making mechanisms relating to teaching from those relating to research. To this end OC&W has introduced programme financing of research in universities, 'research schools' for advanced PhD education, and evaluation procedures of the outcomes of research projects. These evaluations are conducted by the universities themselves (through the VSNU). Reallocation of funds between universities remains difficult because it is determined by law that in the Netherlands all universities should have equal opportunities. Reallocation within universities is more successful as evaluations have provided boards of universities with the necessary arguments to move funds around.

THE S&T WHITE PAPER 'KNOWLEDGE IN ACTION'

In June 1995 a major S&T White Paper, *Knowledge in Action* (OC&W, EZ and LNV, 1995), was published. It set out the course for S&T policy for the coming years and proposed a number of new policy initiatives. It was submitted by the Council of Ministers to Parliament bearing the signatures of the Ministers of OC&W, EZ and LNV together. The new initiatives were presented along three lines: creating more favourable basic conditions for innovation; improving the match of demand for and supply of technology and knowledge; and signalling and initiating new opportunities. In financial terms the first entailed the largest and the third the smallest commitment of funds.

The first line – basic conditions for innovative activities – involved issues that affect both the private and the public sector. A number of policy measures provide extra incentives to increase investments in R&D. The tax incentives of WBSO (Law on Stimulating R&D) have been increased and have been supplemented

by other tax and depreciation facilities. TOK (Technological Development Credit) has been made more accessible to SMEs and
extra support for high technology start-ups has been announced.
PBTS (Programmes for Business Oriented Technology Stimulus)
has been merged with other programmes and the prerequisite
that projects should fall within a specific technological area (information technology, materials technology, biotechnology or
environmental technology) has been dropped. This reflects both
the perceived importance of multidisciplinarity and of collaboration between firms and the conviction that industry is in a better
position to choose which technologies are promising enough to
develop. Other proposals were aimed at increasing efficiency
and transparency of the public knowledge infrastructure. One
idea was to increase the independence of universities, establishing arm's-length relationships between ministries and universities
regarding such issues as labour contracts.

The second line of policy aimed at improving the match of demand and supply in the market for knowledge and technology.
Initiatives covered areas such as the establishment of common
priorities in R&D between public institutions and private industry,
R&D networks, and knowledge diffusion. In this context, research
programmes in public institutions are supported by government
funds on the condition that their design takes account of user
needs. Public intermediary organizations co-ordinating basic research are to develop strategic programmes with the explicit
objective of strengthening the innovative capacity of Dutch business. In particular, more input and commitment is expected from
the private sector in the design of educational programmes. Also
there are policies to increase the accessibility to SMEs of the
public knowledge infrastructure. To improve user–producer co-
operation, projects between firms themselves and between groups
of firms and knowledge institutions receive support for the development of strategic innovations.

The third line of policy aimed at signalling and initiating innovative projects. The main approaches are to signal new opportunities
with high future potential that carry high risk, removing constraints, for example in the sphere of regulation, which could
block promising developments and using the state's position as
both a producer and a user of information and knowledge in
specific areas. Technology policy concerns may lead to govern-

ment procurement of innovative products and services in order to handle problems in fields such as health care, care for the elderly, the fight against crime, infrastructure construction and maintenance. Major new fields in which special initiatives were announced are the development and exploitation of electronic highways (with the ambition to make the Netherlands one of the top three IT service providers in Europe) and the development of environmentally friendly technologies with high market potential (the EET programme, for which topics such as industrial water consumption, industrial waste, environment friendly product development and reduction of traffic emissions have been choosen).

The Ministry of Economic Affairs has a range of policy instruments to stimulate R&D in enterprises. As of 1994 fiscal policies are an important part of Dutch science and technology policy: for example, a partial tax break on the costs of labour employed in R&D (WBSO). Another instrument to support technological development is the set of rules for depreciation of intangible assets transferred from abroad to the Netherlands, which are quite lenient. This policy should also stimulate foreign direct investment into the Netherlands. In an attempt to prevent companies from moving their R&D facilities to other countries as well as to attract investments in new R&D facilities, the regulations governing the depreciation of laboratory buildings have also been relaxed. The dangers of the use of tax instruments are generally recognized. Tax facilities can lead to distortions of the tax system, create loopholes, and as an instrument are difficult to target. A lot of effort has been invested in the construction of the facilities and severe sanctions are imposed upon abuse. Currently, stimulating R&D in firms has such a high priority in government that the Ministry of Finance has lent support to this mechanism.

The Business Oriented Technological Co-operation Scheme (BTS) subsidizes company-orientated co-operative technological research with a particular focus on multidisciplinary and collaborative research. The purpose is not only to stimulate R&D itself, but also to increase the effect of the R&D investments by accelerating the diffusion of the results (by collaboration between companies and by collaboration between companies and research institutes). To receive a subsidy under the BTS policy, projects must be collaborative endeavours with at least two partners, and they must be at least 50 per cent concerned with basic and industrial research. Tenders are submitted for subsidies and support goes up to 37.5 per cent of project costs. The BTS scheme replaced a number of previous support programmes targeted at specific technologies (biotechnology, new materials, environmental technologies and IT). A fund for innovative research programmes (IOPs), to be carried out in the science base to the benefit of industry, complements the BTS scheme. The IOP scheme funds research at universities in one of the areas of the national technology programmes. Programmes exist or have existed on the following themes: materials; information technology; biotechnology; and environmental technology. Through IOPs the government stimulates collaborative research between companies, research institutes and universities. Programmes last eight years and should establish an enduring network between industry and academia.

Among the other financial policy instruments of EZ are a credit facility for innovative development projects: the Technical Development Credit (TOK), which offers loans that cover 25–40 per cent of the costs of an R&D project (including pilot studies). Interest is payable and the loans have to be repaid (in ten years at most) if the project is successful. In the past, 68 per cent of the loans were repaid. TOK aims at stimulating the development of new products, processes and services and is available to companies with at most 20 000 employees world-wide.

TECHNOLOGICAL AID FOR SMEs

The 18 Innovation Centres (ICs) throughout the Netherlands provide services to Dutch SMEs in manufacturing, especially in the metal/electronics, construction, transport and other related sectors. In 1994, almost half of the advisory activities were carried out in the metals and electronics sectors. The service package offered by the innovation centres consists of: providing information and basic advice to SMEs with the possibility of referring them to 'expert' organizations; providing guidance on individual or collaborative technological projects; doing rudimentary company scans (with the option of entering into a guided transformation process, which must be paid for); and helping with commercialization of inventions.

In 1994, the advisory activities of the ICs were mainly concerned with product development (42 per cent), production improvement (27 per cent), and management (31 per cent). In 1997 the Innovation Centres merged with the Centre for Microelectronics and in 1998 with the regional offices of IMK, another institute working for SME's. This was done to create more transparency for the SMEs and in order to provide an all-round package of advisory services, comprising all aspects of entrepreneurship ranging from technical and legal to organizational and managerial. The ICs try to stimulate the building of clusters of companies and of networks between companies and knowledge-producing organizations. In 1998 the Ministry of Economic Affairs was budgeting to spend approximately 70 million guilders on the Innovation Centres.

Besides the ICs there are also policies targeted explicitly at meeting the needs of SMEs. There is the KIM scheme (Knowledge Carriers in SMEs), under which the Ministry of Economic Affairs pays 50 per cent of the salary of young technical graduates who are employed in low-tech firms of fewer than 50 employees to carry out innovative projects within the firms for one year. The programme is managed by the Innovation Centres. This scheme has proved very successful, leading to exhaustion of the budget in the pilot stage. Recently, KIM has been extended for another three years. The KIM instrument is complemented by other schemes through which students of various types (those doing a PhD, a degree in industrial design or the last year of a polytechnic education) can be hired by SMEs or work on problems put forward by them.

Another policy of the Ministry of Economic Affairs to support SMEs is the feasibility allowance. It stimulates diffusion among technological followers by financing feasibility studies to help them to become acquainted faster with new technologies. This measure also proved a success: its budget was exhausted within six months after its launch, suggesting that it certainly meets a specific need of non-front runner SMEs. This scheme has also been extended.

An important stimulus to technological development in business has been the programmes of ICES-KIS (Interdepartmental Commission on Economic Structural Policy/Knowledge Infrastructure). Over the period 1994–8, ICES-KIS spent 250 million guilders on technology and innovation projects that were carried out as public–private partnerships. The funds for ICES come from the sale of natural gas by the Dutch state. There were projects in eight fields: high-performance computing and networking; information technology and civil technique; subterranean construction; ground decontamination; biotechnology; transport and logistics; the development of Mainport Rotterdam; and agricultural production chain management. Project selection and administration in each field was entrusted to a foundation in which government and business co-operated. ICES has provided the Netherlands with important experiences in public–private co-operation and co-financing of R&D. A new ICES-KIS initiative is being developed for the period 1999–2002.

In the Netherlands, policy makers put a relatively heavy emphasis on stimulating co-operation (among firms and between firms and institutes in the science base) as a way to improve performance in technological development and innovation. The necessity to do so is supported with arguments about greater cost-effectiveness and about being a small country opening up ever more to global competition, and it is illustrated with figures about lags relative to competitor nations in the private sector funding of research carried out in the science base (Ministry of Economic Affairs, 1997). The main instruments are of a financial nature. However, there is a growing awareness that the business community also needs other types of government support that are complementary to subsidies. An important avenue is active support in bringing together partners (firms and research institutes) in a co-operative project, for example by developing and maintaining an automated system with a database of technological development activities and potential partners, and organizing workshops on technological co-operation. Another important type of support is help in forging legal contracts to formalize co-operation in R&D (Bureau Bartels, 1997).

Generally the importance of 'org-ware', the organization and management of enterprises and of production chains, for getting technological development going and making innovation pay off, though widely recognized, is not yet reflected in policy that much. Though there are signs that modern production methods and organizational concepts such as concurrent engineering, rapid prototyping and production chain integration are only slowly picked up by Dutch firms, there are few policy initiatives which tackle the non-material side of innovation. The emphasis on benchmarking in the UK, with facilities such as Management Best Practice support initiatives, has no counterpart on that scale in the Netherlands (so far).

3.5 POLICIES TOWARDS EUROPE

Structures and Attitudes

The mechanism by which the Netherlands tries to exert an influence on proposals as they are developed in Brussels and by which decisions are taken regarding the national position *vis-à-vis* policy proposals coming from Brussels is rather complex. As regards the European Commission, useful channels are members of the Dutch Permanent Representation and Dutch officials in strategic positions within the European bureaucracy. At the moment informal contacts and information exchange between Brussels and The Hague are at a satisfactory level, especially in the lower hierarchical tiers. However, hardly any formal structures are in place: there is, for example, no systematic gathering of relevant information from Dutch lobbyists active in Brussels. Concerns are sometimes voiced about the Dutch representation at the higher levels in the hierarchy; too few high-ranking Dutch government officials dealing with technology policy travel to Brussels. To a large degree, things seem to depend upon the vision and priorities of individual top-level civil servants in Dutch ministries.

In the important CREST Committee, the Netherlands is represented by one civil servant from OC&W and by one from EZ. It is currently the policy of the Dutch government to send as delegates to such Committees persons who can not only effectively represent Dutch interests, but who also bear some responsibility for national policy making. There is increasing attention to the legitimation of Dutch positions at these meetings. Nevertheless, there is a feeling in, for example, the AWT that the Dutch representation in Brussels, in particular through CREST, is not high powered enough and that Brussels deserves a higher place on the national R&D policy agenda.

In the informal procedures of the Commission, the Netherlands negotiates in a fairly low-key way. If a major national interest is perceived, active lobbying takes place in a rather *ad hoc* way. It is felt that the Netherlands takes too little advantage of exploiting links between what happens in different organizations (the Commission, the European Parliament and the Council of Ministers).

Dutch influence regarding R&D in the European Parliament (EP) is exercised mainly through Dutch MEPs who are members of the Committee for Energy, Research and Technology (CERT). The Netherlands is not very active in exploiting this line of influence. The Dutch Permanent Representation has very limited manpower available to monitor the EP and officials seldom attend committee meetings. Contacts with Dutch MEPs are of an informal nature. However, since the increase in powers of the EP through the introduction of co-decision making, the attitude of the Dutch bureaucracy towards the EP seems to be changing.

The Ministry of Economic Affairs is responsible for the preparation of the Dutch contribution to the Council of Ministers on S&T policy. All ministries with an interest in S&T policy contribute to the preparation of the Dutch position. The Minister of EZ, in the capacity as co-ordinating minister for technology policy, is the first representative of the Netherlands in the Council of Ministers of the EU, but alternates attendance with the Minister of OC&W, depending on the topics on the agenda.

Generally an official of the Dutch Permanent Representation in Brussels attends the meetings of the Committee of Permanent Representatives (COREPER) and the working groups that figure below COREPER. This official receives instructions from The Hague. The Ministry of Foreign Affairs is in charge of the co-ordination of Instruction Meetings to convey the Dutch position on policy issues to the Permanent Representation in Brussels. EZ is represented in this Instruction Meeting by its Europe Directorate.

The final decisions on the position of the Dutch government in the EU Council of Ministers itself are taken by the Dutch Cabinet. The Co-ordination Committee for European Integration and Association (CoCo) is a committee at official level with representatives from all departments. Its brief is to prepare the position for the Council for European Affairs, a subcouncil of the Dutch Cabinet. CoCo is chaired by a Junior Minister of Foreign Affairs. Here too, EZ is represented by its Europe Directorate. Consequently, although technology policy on the national scene is the responsibility of the Directorate for Technology Policy of EZ, this Directorate does not deal directly with proposals on technology policy at the European level. Here it is represented by the Europe Directorate which acts through a Council under the control of the Ministry of Foreign Affairs.

Defending Dutch interests in the Council of Ministers is considered the most important route to influence EU policy. However, to influence the discussions effectively it is vital to have all the relevant information at an early stage, to establish a clear strategic position early, and to have a very able and highly qualified negotiating team that has been granted sufficient autonomy and flexibility. The Dutch formal structure is badly adjusted to fulfil these demands. The requirements to operate effectively in Brussels do not really match the slow bargaining and negotiation culture of Dutch policy, with its emphasis on compromises, coalitions and conflict avoidance. Consequently officials in ministries often try to bypass formal circuits, the articulation of Dutch interests in Brussels is often chaotic and the positions taken are frequently rather late and of an ambiguous nature.

Budgets

Unlike in the UK, the Netherlands has not adopted a compensation mechanism to allocate Dutch contributions to the EU to chapters in the national budget. National policy is seen as complementary and preparatory to European initiatives. National technology policy is aimed at knowledge exploitation and technology diffusion, often closer to the market and more accessible to SMEs. It is generally seen as unwise to decrease national budgets in response to a rise in EU budgets, especially as R&D budgets in competitor nations are rising. EU budgets are seen as additional because EU policies do not overlap with but are complementary to national policies. Moreover, the Netherlands will only be able to attract EU funds and take advantage of collaborative research projects if national efforts are maintained at a sufficiently high level.

3.6 CONCLUDING OBSERVATIONS

Across the board, 'philosophies' on R&D policy in the Netherlands tend to be rather pragmatic. Policy makers do not seem to be particularly interested in economic or political theories or ideologies. Policy proposals are developed inductively in response to perceived trends and international comparisons and with an eye on what happens in competitor nations. At the moment there is a fairly broad consensus concerning the central objectives and instruments of technology policy. Clearly the ball is increasingly put in the hands of business enterprises: they are expected to indicate what technologies should be developed, how conditions for innovation should be improved, and what knowledge and skills the educational system should provide in the future. Government should support industry by creating favourable conditions but also by providing access to all possible sources of information and technology. Although there is general agreement on the way ahead, a source of tension exists in the division of responsibilities. Government points to the fact that industry has a major interest in the policies pursued and strongly influences the ways in which money is spent. It should therefore bear its snare of the cost of the policy. Industry tries to limit its commitments, arguing that maintaining the knowledge infrastructure is the responsibility of the state.

Issues of technology policy in the Netherlands are discussed in a fairly broad context. There is attention to technological progress as the key to competitiveness and wealth creation. However, technology as an instrument to solve society-wide problems, to relieve pressures on the environment, as having a cultural value in itself is also usually part of the debate. There is a relatively large emphasis in Dutch technology policy on issues of public awareness and acceptance of technology. Also, considerable funds are spent on technology assessment, the systematic assessment of the impact of specific technological developments on society from an economic, social, environmental, ethical, political or other relevant point of view.

An overriding concern driving current Dutch R&D policy is the attractiveness of the Netherlands as a location in which to invest in economic activity. Policy makers are acutely aware of the openness of the Dutch economy, and thus of both the danger of domestic firms relocating abroad and the opportunity that openness implies for attracting foreign firms to the Netherlands. The Netherlands seems very successful at attracting certain functions within foreign firms, especially distribution and marketing. A central policy concern is to create conditions that stimulate firms to invest in the Netherlands not only in distribution and marketing activities but in higher value-added activities such as R&D as well. The idea of policy competition among European governments is a common thought among Dutch policy makers. Ensuring a level playing field for industry is an important argument in the formulation of R&D policy.

A recent trend in support for R&D in the Netherlands is towards more generic policy instruments. In contrast to the UK, tax policies are seen as appropriate because they are of a generic nature and provide relatively easy access for all firms. This reflects the influence of employers' organizations on policy formulation, who have been lobbying for this approach, as well as a retreating government which has been more and more prepared to leave decisions on the content of R&D projects to the market. Another important trend is the creation of arm's-length relations within the knowledge infrastructure. Both universities and public research organizations are made ever more independent of ministries, both freedom of action and responsibility for finance being transferred to these institutes.

Current support policy is directed to a large extent towards SMEs. However, SMEs themselves are relatively marginally involved in policy design; the initiative here lies clearly on the side of government. Increasing attention is also being given to larger non-multinational firms. While it is being felt that the large Dutch multinationals are able to take care of themselves, or at least that the government is not able to raise enough funds to help them effectively, it is recognized that non-multinational larger exporting firms with the Netherlands as a home base (for example, DAF, Océ) are important assets to the Dutch economy and may need access to support under specific circumstances.

REFERENCES

Adviesraad voor het Wetenschaps- en Technologiebeleid, 1994, *Technologiebeleid en economische Structuur*, Den Haag.
Adviesraad voor het Wetenschaps- en Technologiebeleid, 1996, *Reactie op het wetenschapsbudget 1997*, Den Haag.
Adviesraad voor het Wetenschaps- en Technologiebeleid, 1998a, *Het nut van grote technologische instituten*, Den Haag.
Adviesraad voor het Wetenschaps- en Technologiebeleid, 1998b, *Reactie op Stategisch Plan TNO 1999–2002*, Den Haag.
Arthur D. Little , 1996, *A Technology Map for Dutch Business*, Den Haag.
Bureau Bartels B.V., 1997, *Evaluatie van de PBTS, IT-regeling en de Clusterprojecten*, Utrecht /Assen.
Bureau Bartels B.V. and Bakkenist Management Consultants, 1996, *Evaluatie van de Wet Bevordering Speur- en Ontwikkelingswerk (WBSO)*, Utrecht /Assen /Voorburg/ Diemen.
Centraal Bureau voor de Statistiek, 1996, *Kennis en economie 1996*, Voorburg/Heerlen.
Centraal Bureau voor de Statistiek, 1997, *Kennis en economie 1997*, Voorburg/Heerlen.
Corvers, F., R. Hassink, M. Slabbers and B. Verspagen, 1994, *Monitoring technology policy in Europe*, Maastricht: MERIT.

Diederen, P.J.M, N. van den Eeden, and J. Kuper 1998, *Economische en bestuurlijke evaluatie van de Bijdrageregeling innovatieprojecten*, Den Haag: Landbouw Economisch Instituut (LEI).

Ministerie van Economische Zaken, 1993, *Concurreren met Kennis*, Den Haag.

Ministerie van Economische Zaken, 1995, *Technology Policy in The Netherlands*, Den Haag.

Ministerie van Economische Zaken, 1997, *Toets op het concurrentievermogen 1997; Klaar voor de toekomst?*, Den Haag.

Ministerie van Landbouw, Natuurbeheer en Visserij, 1995, *LNV-Kennisbeleid tot 1999*, Den Haag.

Ministerie van Onderwijs, Cultuur en Wetenschappen, 1992, *Speur- en ontwikkelingswerk in Nederland; Beleid, financiering en uitvoering*, Zoetermeer.

Ministerie van Onderwijs, Cultuur en Wetenschappen, 1994, *Wetenschapsbudget 1995*, Zoetermeer.

Ministerie van Onderwijs, Cultuur en Wetenschappen, 1995, *Onderzoek in cijfers 1995*, Zoetermeer.

Ministerie van Onderwijs, Cultuur en Wetenschappen, 1996, *Wetenschapsbudget 1997*, Zoetermeer.

Ministerie van Onderwijs, Cultuur en Wetenschappen, 1997a, *Voortgangsrapportage wetenschapsbeleid*, Zoetermeer.

Ministerie van Onderwijs, Cultuur en Wetenschappen, 1997b, *OC&W in kerncijfers 1998*, Zoetermeer.

Ministerie van Onderwijs, Cultuur en Wetenschappen, Ministerie van Landbouw, Natuurbeheer en Visserij, Ministerie van Economische Zaken, 1995, *Kennis in beweging*, Den Haag.

Minne, B. 1995, *Onderzoek, ontwikkeling en andere immateriële investeringen in Nederland*, Den Haag: Centraal Planbureau (CPB), research memorandum.

OECD 1987, *Reviews of National Science and Technology Policy; The Netherlands*, Paris.

Overlegcommissie Verkenningen, 1994, *Koersen op kennis*, Amsterdam.

Overlegcommissie Verkenningen, 1996, *Een vitaal kennissysteem; Nederlands onderzoek in toekomstig perspectief*, Amsterdam.

Rademaker, E.H., 1994, *Europese integratie en nationaal technologiebeleid*, Leiden/Rotterdam, Msc-thesis.

RAND Europe, Coopers and Lybrand, Innovation and Technology SA (Batelle), 1998, *Technology radar*, Den Haag: Ministerie van Economische Zaken.

Slabbers, M. and B. Verspagen, 1995, *Stemming 2, De Nederlandse technologische positie en de invloed van globalisering*, Maastricht: MERIT.

Sociaal-Economische Raad, 1995, *Kennis en Economie*, Den Haag.

Tijssen, R.J.W., Th.N. van Leeuwen, B. Verspagen and M. Slabbers, 1994, *Het Nederlands Observatorium van Wetenschap en technologie*, Leiden: CWTS and Maastricht: MERIT.

Werkgroep Markt en Overheid ('Commissie Cohen'), 1997, Eindrapport Project Marktwerking en Wetgevingskwaliteiten, Den Haag.

Zwetsloot, F.J.M., ed., 1997, *Sturing van wetenschappelijk onderzoek*, Den Haag: Delwell.

4. France

4.1 INTRODUCTION

France has a strong science base and a solid tradition in research. It spends more on R&D as a percentage of GDP (2.34 per cent in 1995) than the UK or Germany. Nevertheless, the French science and technology system is troubled by a sense of crisis. The general feeling seems to be that a major revision of the system is needed; the current structures are not delivering what is expected from them. Changes that have been implemented, especially those that took place in the early 1980s and the current moves towards decentralization, have not sufficiently improved the situation. A more fundamental transformation is long overdue but is, as everywhere, fiercely resisted by vested interests.

The sense of crisis and the need for change which is felt with regard to the science and technology base is not an isolated phenomenon. Overall there is an increasing tension in the relationship between the state and the economic system. Traditionally the French economic system has been driven forward by the state and a benevolent elite. Ties between the state and the business sector have always been manifold and tight, both in terms of ownership and of personal links. Many of the largest French companies were until recently or still are in the hands of the state. Examples are Crédit Lyonnais and Banque Paribas in banking, Renault in vehicle production, GEC-Alsthom in the manufacture of power stations and transport equipment, Alcatel in communication, Thomson in electrical equipment, Air France as the national airline, and Total and Elf-Acquitaine as oil companies. More than in other developed countries, managers seem to switch back and forth between these state-owned companies and public administration. Thus the French government is directly involved in the management of production for the market.

One of the consequences of having this type of 'state capitalism' is that every economic conflict has direct political repercussions. Close links between the business and government sectors lead to an entanglement of private and public interests and responsibilities. In times of economic adversity, labour unions and employers' organizations are soon inclined to look to the government for help to solve their conflicts. Large state-owned companies effectively operate under a 'soft budget constraint' and interest groups in both camps, on the management and on the workers' sides, exploit this situation.

Air France, Bull Computers and Crédit Lyonnais provide the most striking examples of what happens when these companies get into deep trouble. The state saved them from bankruptcy by the injection of enormous amounts of taxpayers' money, keeping them temporarily afloat in an ever more competitive market environment. Though state intervention can rescue firms from immediate financial breakdown, it usually falls short of reorganizing them to a sufficient extent, or even hampers the management in doing so. Generally, state support does not inject enough entrepreneurial spirit into old structures to make them survive in the long term.

The left-wing government under Lionel Jospin recognizes too much state involvement in business as one of its fundamental sources of trouble and, also under the pressure of European regulation which limits support to companies by national governments, has embarked on a path of reform and privatization. This process evolves slowly in order not to lose the support of the communist coalition partners and risk unmanageable conflicts with radical labour unions. Meanwhile, the political opposition is involved in its own troubles. Internal fights about political strategy and about collaboration with the extreme right nationalists have divided the political right.

One factor that aggravates the current sense of crisis is the economic situation of persistent unemployment, modest growth and the need to decrease the public deficit which renders any increases in public spending politically impossible. This has a direct impact on any type of public spending, and therefore also affects science and technology policy. Due to the structure of French public R&D spending, which flows to a large extent through public research institutes, the room for manoeuvre, given a static budget, is rapidly disappearing. France conducts a large share of its research in government laboratories, the staff of which enjoy civil servant status. Given the age structure (the age of researchers is 47.5 years on average, ten years older than in the UK) and the low turnover rate of this labour force, wage-related expenses are high and grow autonomously (according to estimates given to us, by 4–5 per cent per year). This inevitably leads to a crowding out of necessary investment in new initiatives.

4.2 STRUCTURES

The French national system of innovation is dominated by a few large technological clusters, 'subsystems of innovation', as Chesnais (1993) calls them. Each subsystem is built around a few large companies that hold close ties with the government. It could be argued that this system of close links between private enterprise and the state has developed in the past as a consequence of a lack of a tradition of private entrepreneurship and a dynamic banking system. Chesnais distinguishes five, partly overlapping subsystems of innovation: military products; nuclear power; telecommunications; space; and aeronautics. A few firms, such as Framatome, GEC-Alsthom, Aerospatiale, Thomson, Matra, Snecma, Dassault and Alcatel dominate the subsystems. The organization of the national system of innovation has traditionally been more directed at fulfilling the needs of these large companies than at supporting SMEs. The military subsystem is relatively large. It competes with the civil sector for capital and labour and, given its substantial size, it may strain the civil sector by its drain on human capital in the national scientific labour force. Disagreement exists as to the extent to which military R&D in fact crowded out civil R&D in the past. Some would argue that military R&D is nearly completely additional; if these funds had not been devoted to subsidies for military research and development in the private sector, they would not have been spent on R&D at all. On the other hand, there have been important spillovers from military research to civil applications in the fields of information technology, telecommunications and aerospace. The size of those, however, is difficult to estimate.

An overview of R&D spending in France is presented in Table 4.1. Public sector spending accounts for about half of the total spend on R&D, which is more than the EU average, the US and Japan. The two main spending cate-

Table 4.1   Overview of R&D spending in France in 1994 (FF billions)

	Total spending	% of total	% of public spending
Public spending			
Fundamental research	27.6	15.5	31.4
Military research	26.2	14.8	29.8
Grand Technological Programmes	14.5	8.2	16.5
R&D for public policy purposes	12.7	7.2	14.4
Support of industrial R&D	4.1	2.3	4.7
EU programmes	2.3	1.3	2.6
Regional policies	0.6	0.3	0.7
Total	88.0	49.5	100
Private spending			
Industrial R&D	89.6	50.5	
Grand total	177.6	100	

Source: OST, *Indicateurs 1998*.

gories in the public part are fundamental research and military R&D, each accounting for almost 30 per cent. Then there are two categories with a share of around 15 per cent of the total budget: R&D expenditures for public policy purposes and investments in the so called Grand Technological Programmes. Finally there are three minor categories: subsidies to support private sector R&D, spending on regional programmes and spending on research programmes of the European Union.

French enterprises often spend a smaller percentage of their turnover on R&D than comparable firms in other countries; this is considered a significant weakness of French R&D. Of the R&D which firms carry out in-house, they pay for less than 75 per cent themselves; on average in the EU and the US this is more than 80 per cent. These figures serve to illustrate the dependence of the private sector on the public R&D effort. The private R&D effort in France is more concentrated in a few sectors than in other industrial countries of comparable size. Two-thirds of expenditures are accounted for by the space and aircraft, electronics, automotive, pharmaceutical and chemical industries (see Table 4.2). What is worrying about this is that these are the

Table 4.2   Industry R&D spending (in-house) and government funding in 1995

	Total R&D expenditures (FF million)	Share of enterprise R&D funded by the government (%)	Share of total state funding of industry R&D (%)
Aeronautics and space	14 416	42.6	39.9
Automotive	14 092	0.8	0.7
Pharmaceutics	13 091	1.9	1.7
Communication equipment	11 996	12.7	9.9
Instruments, other electronic equipment	11 302	35.0	25.7
Chemicals	6 757	6.4	2.8
Machines and equipment	5 299	20.3	7.0
Energy	4 179	4.3	1.2
Computers	2 922	20.4	3.9
Other	25 159	4.4	7.2
Total	109 213	14.1	100

Source: *Projet de Loi de Finances 1998.*

industries of today, not those of tomorrow. France is relatively weak in emerging technologies such as information technologies and biotechnology. State support for private R&D primarily flows to the aeronautics and space and the electronic equipment industries. This reflects the importance of public funding of defence R&D. Sectors that show a more than average dependence on the state for their R&D budgets are, in addition to the ones mentioned, the computer industry, the communication equipment industry, and other machinery and equipment industries.

The French public science and technology structures are at the same time both complicated and simple (see Figure 4.1). They are complicated in that there seem to be many layers and participants (there are in all 23 public sector research organizations and 160 institutions of higher education), and simple in that one ministry, the Ministry of Education, Research and Technology (*Ministère de l'Education Nationale, de la Recherche et de la Technologie*, MENRT), dominates the civil R&D field. There are separate civil and military structures, though there is some overlap and the division between the two is not very sharp.

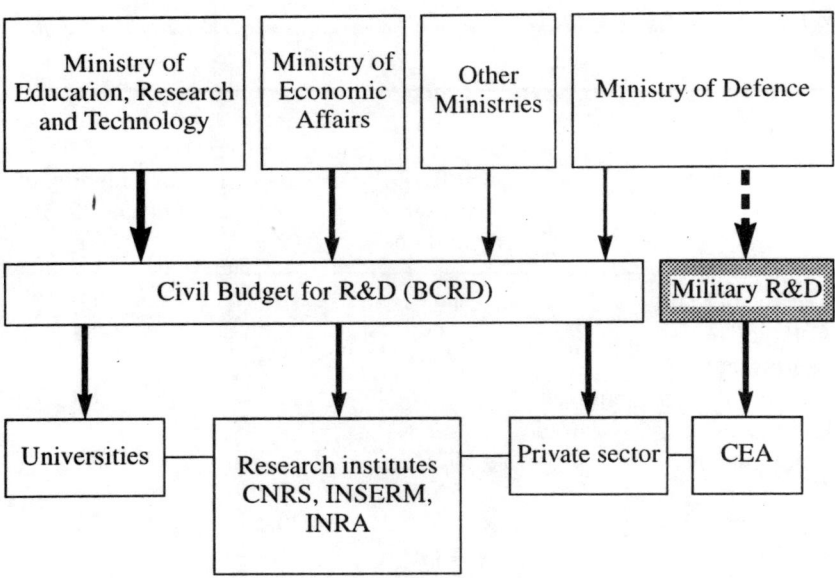

Figure 4.1   The structure of the French public system of science and technology

Ministries

The Ministry of Education, Research and Technology, by far the most important ministry in this field, has two functions. First, it is the co-ordinating ministry for civil R&D and is responsible for putting together the civil R&D budget (*Budget Civil de Recherche et de Développement technologique*, BCRD). Second, it is responsible for the larger share of public civil spending on science and technology (see Table 4.3). In its first function its position is very similar to that of the German Ministry of Education, Science, Technology and Research (BMBF) before it was reorganized. The BCRD is an inter-ministerial R&D budget that the Ministry of Education, Research and Technology negotiates with other ministries. It covers all civil R&D expenditures by the central government. The BCRD was created in order to enhance co-ordination in the R&D field between spending ministries, the lack of which is acknowledged to be a problem in other case study countries, such as Finland. When agreement among spending departments has been reached, MENRT negotiates the total with the Treasury. The role of the Treasury is similar to that in other countries. Frame-budgeting, whereby the Treasury agrees on the grand total of spending for a ministry but leaves decisions on details about how the money is spent to spending departments, is applied in France as well. A provisional BCRD, drawn up under the co-ordination of MENRT, is usually discussed in Parliament in the autumn. Generally mem-

Table 4.3   *Civil R&D expenditures per ministry in 1998:* dépenses ordinaires *plus* crédit de paiement[a]

Ministry	Expenditure (FF million)	Share (%)
Education, Research and Technology: university research	2 362	4.5
Education, Research and Technology: research institutes	39 611	74.7
Economic Affairs, Finance and Industry	6 367	12.0
Equipment and Transport	2 132	4.0
Other ministries	2 583	4.9
Total R&D expenditures	53 055	100

Note: [a] *dépenses ordinaires*: basically salaries; *crédit de paiement:* amounts that can be spent in a particular year.

Source: *Projet de Loi de Finances.*

bers of parliament look favourably upon R&D matters. However, when it comes down to operational matters, that is, how the money is spent, Parliament has little influence.

CIVIL BUDGET FOR RESEARCH AND DEVELOPMENT (BCRD)

The BCRD, the total civil R&D budget, was originally created to foster co-ordination and co-operation between spending ministries in matters of R&D – the lack of which is acknowledged to be a problem in other case study countries such as Finland. This seems to have worked in years of rapid growth of public R&D spending. The situation was especially to the advantage of the Ministry of Education, Research and Technology (MENRT) as it was held responsible for the BCRD. Other ministries are in a sense subordinate to MENRT with regard to R&D expenditures. When the BCRD started to decline in real terms, it became evident that the BCRD does not really foster co-operation and might even harm it. Other ministries feel it is not in their interest to exert a lot of effort to support the BCRD as it is not their direct responsibility and all the acclaim for any positive results tend to go to the MENRT. A side-effect of the BCRD is that is seems to provide cover for small R&D budgets and programme proposals in times of tight funding. Exposed to parliamentary examination, and especially to the scrutiny of the Treasury, they might have fallen victim to budget cuts at an earlier stage. Now they are treated as an integral part of the BCRD in Parliamentary discussions.

The Ministry of Economic Affairs, Finance and Industry (MEFI) is the second-largest spender on civil R&D. Though clearly less prominent than MENRT, it still has an important position in the French public R&D system. It is responsible for connections with industry and for support for privately conducted R&D. MEFI funds space research via the *Centre National d'Etudes Spatiales* (CNES). Moreover, most government finance for civil research conducted by industry is channelled via the Ministry of Industry. Finally, the *Agence Nationale de Valorisation de la Recherche* (ANVAR) is part of the Ministry. ANVAR's main task is to facilitate the transfer of technology from government laboratories to industry.

The Ministry of Defence plays a role in R&D that reflects the weight and importance of the arms industry in France. Military research accounted for 29 per cent of public spending on R&D in 1996, down from around 37 per cent ten years ago. This is still below the UK (37 per cent in 1996) and the US (55 per cent in the same year), but in absolute numbers the UK spends less on military R&D. Military research can be divided into three parts: nuclear R&D undertaken by the *Commissariat à l'Energie Atomique* (CEA), R&D carried out by state-owned traditional arms and munitions manufacturing companies, and R&D commissioned to private sector industrial firms (Chesnais, 1993). It should be noted, however, that the distinction between civil and military R&D in France is less clear than it is in Germany, for example. The Ministry of Defence channels some of its budget through the budget for civil R&D. The combination of civil and military R&D takes place mainly within the CEA, which conducts research on both nuclear energy and nuclear armaments, and within the space programme. In France, just as in the UK, defence R&D financed by government is largely undertaken by industry. In France about 45 per cent is carried out by industry (OST, 1998), in the UK almost 70 per cent. Closely linked with the Ministry of Defence is the *Délégation Générale pour l'Armement* (DGA). DGA is in charge of arms procurement, most of which takes place without tendering. It is alleged to be one of the strongest drivers of French R&D (Chesnais, 1993).

The main objectives of the Ministry of Defence are the establishment of a unified European defence industry that can compete with the US defence industry, and the reduction of armament acquisition costs. With this context in mind, projects are selected according to technological and military criteria. In 1994, 83 per cent of public R&D funds handed to the private sector flowed through the ten largest French industrial conglomerates linked to the defence establishment which carry out about one-third of French private sector R&D (Guillaume, 1998). Private industrial partners will become, even more than in the past, main contractors. These principal contractors can then subcontract institutes in the public knowledge base for research input. It is hoped that organizing military research in this way will increase the efficiency of public defence research, as it improves the transformation of research results into useful and perhaps commercially attractive weaponry (*Projet de Loi Finances*, 1998).

In France, the Ministry of Transport plays a significant role in R&D; in this respect France differs from our other case study countries. Its importance is solely dependent on the large share of aviation-related R&D, the civil part of which is channelled via the Ministry of Transport. All other spending ministries play a minor role in funding R&D, but have some influence on R&D carried out in their respective policy areas. The Interministerial Committee on Scientific Research and Technology (CIRST), formally headed by the Prime Minister, is a way to co-ordinate the policy efforts of different ministries.

Research Institutes

Most French government research organizations are categorized as belonging either to the *Etablissement Public à caractère Scientifique et Technologique* (EPST) or to the *Etablissement Public à caractère Industriel et Commercial* (EPIC). However, the difference between these categories is not very large; differences depend on habits and style of management, making the vaiation within each category as large as the inter-category variation. EPST institutes were created in the early 1980s mainly by changing the status of existing government laboratories. Their new status empowered them with more freedom to determine their own strategies, for example, by giving them the right to establish subsidiaries, some of which since have been launched on Nasdaq, the American high-tech stock exchange. Most research conducted on the initiative of government takes place in EPST institutes. EPIC institutes are allowed (and demanded) to engage in some commercial activity. Some of the EPIC organizations relate to the Ministry of Industry. Table 4.4 gives an overview of government funding plans of EPST and EPIC institutes for 1998. There are a number of research institutes operating outside the EPST and EPIC umbrellas, also partly funded by MENRT. The largest and best-known of these is the *Institut Pasteur de Paris*, funded by MENRT with about FF378 million, which carries out research on contagious diseases. ANRS, funded with FF231 million, carries out research on Aids.

The most important EPST institute, consuming around two-thirds of the EPST budget, is the *Centre National de la Recherche Scientifique* (CNRS). CNRS was founded in 1939 with the initial brief to develop and co-ordinate all French science. The role it plays has been likened to that of the German Max Planck Society: the CNRS also focuses on fundamental and strategic research and operates through independent laboratories. Its existence is partly a response to the weak standing of research in French universities. CNRS laboratories are often found in the vicinity of universities.

Of the EPIC institutes, CEA is the largest. Although CEA was originally established to carry out research on the military and energy applications of nuclear technology, it extended the scope of its activities to all kinds of applications of nuclear physics, especially medical applications. Other areas are covered by CEA as well, such as astrophysics and climatology. Like other EPIC institutes, CEA collaborates with ANVAR, the transfer agency, to facilitate technology transfer to the private sector.

The *Centre National d'Etudes Spatiales* (CNES) is the second-largest EPIC institute, far smaller than the CEA. CNES's budget includes French contributions to the European Space Agency (ESA). A large part of its research is contracted out. For those two reasons, CNES's budget is relatively small. Its

Table 4.4   Civil R&D budgets of EPST and EPIC institutes

Name	Description	Budget 1998 (FF million)	Change 1996–8 (%)	Employment
CNRS	General scientific research	13 722	3	26 277
INSERM	Medical research	2 563	4	4 960
INRA	Agricultural research	3 425	3	8 515
ORSTOM	Research on the development of developing countries	1 031	–2	1 609
Other EPST		1 017	5	1 919
Total EPST		21 758	2	43 280
CEA	Research on nuclear energy (and armament, not in BCRD)	6 483	3	11 406
CNES	Research on aerospace	9 065	–2	2 416
ANVAR		923	–16	365
INFREMER	Research on oceans and seas	960	2	1 330
Other EPIC		893	–1	1 933
Total EPIC		2 965	–1	18 324

Source: *Projet de Loi de Finances pour 1998.*

main field of research is the development of satellites and of the rockets necessary to take them into space.

ANVAR is also an EPIC institute. It manages a portfolio of patents and seeks industrial partners for university laboratories and for the CEA, the CNRS and other research institutes. ANVAR also financially stimulates innovative activities of SMEs, supports independent inventors and mediates in the development of innovative networks.

Universities

Higher education institutions in France are either universities or so-called *grandes écoles*. Most French government-funded research is carried out in research institutes with higher education institutions playing a subsidiary role. Research and higher education are therefore in France more separate than in any of the other case study countries. Higher education is the responsibility of the central government (MENRT), whereas high school education, up to the *baccalauréat*, is the responsibility of local government (the 22 regions). The *grandes écoles* are the elitist institutions which educate the *grand corps* which runs much of France. Top functions in public administration and industry are often only attainable by those who graduate from these 'great schools'. The two most important among them are the *Ecole Nationale d'Administration* (ENA) and *Ecole Polytechnique*. The *grandes écoles*, the predecessors of which were established in the Renaissance, are deeply rooted in the tradition of French higher education. There are 238 *écoles d'ingénieurs* and 230 *écoles de commerce*, educating together just under 10 per cent of French students (that is, those who enrol in tertiary education). The universities take care of about 62 per cent of students and the system of technical tertiary education teaches another 16 per cent. The universities have grown much faster since the beginning of the century than the *écoles*.

DIFFERENCES BETWEEN THE *GRANDES ÉCOLES* AND THE UNIVERSITIES

Universities and *grandes écoles* operate under different regimes and with distinct objectives. They used to differ in the following ways:

- The main purpose of university teaching is to educate people to become scientific researchers. The *grandes écoles* provide top level vocational education for future civil servants and managers.
- Universities are legally obliged to accept anybody with a certificate of secondary education. To enter a *grande école* students have to go through an arduous selection procedure.
- Similar degrees of different universities (except doctorates) are supposed to be of equal value. Among the *grandes écoles* there is a hierarchy.
- Universities have an administrative structure in which the director (*président*) shares authority with researchers, students and MENRT. The *grandes écoles* are managed by a powerful board of directors.
- The working facilities (such as libraries, restaurants, sport) and living conditions offered to students of the *grandes écoles* are far better than those offered to university students.

- Research is a principal activity in universities. In the *grandes écoles*, however, although there are excellent research facilities, the relation between education and research is very weak.
- University education is practically free for students. While in some *grandes écoles* students have to pay a fee, students receive money for studying at some of the others.

Regarding the first three points mentioned above, differences have been fading over recent years. Universities are increasingly becoming orientated towards vocational education; they do select their students in one way or another; and the labour market differentiates between their diplomas. As far as the last four points are concerned, the differences between universities and *grandes écoles* are still large. The two types of institution are slowly converging, but the two systems are as yet by no means attuned to each other. The roles and positions of universities and *grandes écoles* are established by law. However, their differences go well beyond legal aspects. For instance, the universities have only half as much money available per student as the *grandes écoles*. Consequently, due to their better financial situation, the *grandes écoles* are able to attract the superior teachers.

(Attali, 1998)

Universities are also charged with the task of facilitating the transfer of research outcomes to the private sector. Each university is obliged to develop structures to support technology transfer and to implement a policy for the exploitation of results adapted to the characteristics and needs of the local region. In the coming years, more emphasis is to be put on the improvement of relations between universities and the private sector, especially SMEs. The government has indicated that universities should not develop stand-alone transfer offices but should actively participate in technology transfer networks. Universities are supposed to stimulate the establishment of innovative enterprises and the mobility of their researchers towards the private sector. These policy objectives point in the same direction as those regarding universities in other countries, but progress seems to have been slower in France than elsewhere. Although universities were charged with this transfer task 16 years ago, the government remains unimpressed with the results so far. Consultancy and service activities have hardly developed and are not very transparent (*Projet de Loi de Finances*, 1998).

Intermediaries

At the regional level, several intermediary structures have developed. Some universities and research institutes have established transfer offices, just like their foreign counterparts have done. On the regional level there are institutes such as the *Centre Régionaux d'Innovation et de Transfert de Technologie* (CRITT). These are joint-venture organizations with private and public financial participation. The CRITT should enhance regional innovation-related networks, arouse interest in new technologies at SMEs, develop innovative processes in collaboration with SMEs and train their personnel. The *Réseaux de diffusion technologique* are linked to ANVAR and other research and intermediary organizations. The *Réseaux* are networks of innovation advisers who should enhance access to and coherence of the organizations which deal with technology transfer. The advisers are not employed at the *Réseaux*, but remain employees of the research institutes and intermediary organizations that participate in the network. The *Réseaux* act as a first-line advisory organization. Advisers pay visits to companies and refer them to other organizations if necessary. One important advantage of the *Réseaux* is that they stimulate the exchange of information between the different organizations participating in the regional innovation structure. The 'one-stop shop' principle makes it easy for SMEs to find their way to the knowledge infrastructure with their questions (Van Riel, 1995).

TECHNOLOGICAL RESOURCE CENTRES (CRT)

Some intermediary organizations are labelled (and certified as) Technological Resource Centres (*Centres de Resources Technologiques* – CRT). The objective of the CRT initiative is to ease the dialogue between enterprises and research organizations and to improve the efforts of those organizations on behalf of SMEs. The idea behind the CRT concept is that effective relations between research organizations, intermediaries and SMEs should be based on professionalism, partnership and competence. The label that is granted to an intermediary guarantees to SMEs that a particular organization fulfils certain criteria with respect to these three issues. Apart from this, each CRT has a clearly defined area of interest and identity and has well-developed links to research organizations. Organizations that are labelled CRT can benefit from financial mechanisms set up to foster the demand of SMEs.

Representation of Private Sector Interests

The close relationship between state and industry ensures a solid representation of private sector interests in government circles. There are several influential government committees on R&D matters in which industry representatives take part. However, in France there does not seem to be an extensive system of consultation on policy matters similar to those in other case study countries. We did not come across this large proliferation of formal consultative procedures and mechanisms for consensus building that commonly supports policy making in the Netherlands and Finland.

Public procurement, which is especially relevant in the military sector, fosters closeness and co-operation between state and industry. But also on the civil side, industry can approach government in an informal way with proposals and plans. Civil servants are regularly in contact with industry to discuss R&D initiatives and possibilities to obtain funding for these. Even if government officials cannot guarantee funding of programmes prior to the setting of the budget, industry seems to get its arguments across to government circles with regard to potential directions. Once the budget is set, the process of applying for funding is relatively straightforward and non-bureaucratic.

4.3   PHILOSOPHIES

The political culture in France is somewhat different from that in countries like Germany, the Netherlands and Finland. There seems to be less of a tendency towards compromise and decision making on the basis of consensus; the inter-ministerial decision making process on R&D (and, presumably, on other issues as well) seems to be based more on competition between departments. French ministers need to show results, and this does not foster co-operation.

Mission Orientation and *Dirigisme*

The French philosophy has always been clearly more on the mission-orientated side of R&D policies than in any of our other case study countries. Government R&D resources tend to be devoted to large-scale projects where the involvement of the state is often substantial. Most public R&D money is spent on space, aerospace, nuclear energy and the telecommunications sectors. French civil servants feel that their main task is to identify direction, then to take special initiatives. The creation of conditions favourable to innovation in the private sector comes last on their list of priorities. In all other case study countries, creating conditions was perceived as the single most important task of the government. This mission orientation is expressed, for instance, by the maintenance of an extensive array of *Grands Programmes Technologiques*, large publicly financed R&D programmes of long duration, which have no rivals on this scale in any other European country. Perhaps the best example of mission orientation is the French nuclear energy programme. As this was partly initiated to produce raw material for the French nuclear armament programme, it also shows the close connection between civil and military objectives.

The Interministerial Committee on Scientific Research and Techniques (CIRST) in 1997 put forward three pillars as the foundation upon which French R&D policies must be based. The first of those is the determination of grand priorities (*une politique scientifique par grandes priorités*). On the basis of three criteria – the recognized strength of the French research base; the potential competitiveness of French industry in a particular area; and the high growth rates of markets – industrial themes have been selected on which research is to be focused. There are four industries that are to benefit espec- ially from these efforts: agribusiness; terrestrial transport and aeronautics; electronics and information technologies; and the chemical industry. Three additional themes have been selected that are less clearly linked to particular sectors: biomedical research; environmental research; and research directed at industrial product and process innovation. This approach illustrates the traditional view which fosters a proactive role of the French state in science and technology matters. It is illustrated in the opening paragraphs of the document in which the Interministerial Committee on Scientific Research and Techniques presents its policies: 'If France succeeded in creating high- tech industries of the highest standard in the world, it did so because it succeeded in developing a clear and ambitious research policy, guided by a concern for *grandeur* and national independence.'

The second pillar supporting R&D policy is sound management of human resources. Two obstacles impeding dynamism in the public research system have been identified: first, the age structure and reward schemes of the labour force which makes the system expensive and will drive up wage costs rapidly over the coming years; second, the very low rate of labour turnover which leads to excessive isolation of the research community. The mobility between the public research infrastructure and the private sector is presently extremely low: no more than about 30 to 40 researchers out of a total of more than 25,000 leave the public system for a job in the private sector, whereas in the US, for example, researchers leaving the public research infrastructure to set up private companies is one of the main drivers of innovative dynamism (Guillaume, 1998). Policies aimed at renewal of the labour force and increas- ing mobility are designed to address these problems. Researchers will be encouraged to move between research institutes and the private sector. Funds have been designated to employ young researchers at public research institutes. The number of positions at research institutes and universities, both for researchers, postdoctorates and support personnel (librarians, technicians, and so on) will be increased (*Projet de Loi de Finances*, 1998).

The third proposed pillar under French R&D policy for the years to come is an increased effort to improve the exploitation of research results for commercial purposes. Various ideas on this topic circulate. Researchers should be rewarded on the basis of their patenting activity as well as their publication record. Knowledge transfer from the public to the private sector should be facilitated, for example by increasing the possibilities for temporary detachment. Enterprises are to gain more influence on the activities of public research institutes, while researchers in the public sector must receive a personal financial interest in the commercial exploitation of the results of their research. One way to do so is to allow researchers to participate in the development of innovating firms, for instance by letting them invest in kind in the form of contributing a patent. Other areas that will receive special attention are the protection of intellectual property and the mobilization of private funds for innovative activities.

Whilst the first pillar is testimony to the fact that mission orientation in the French public R&D policy is still strong compared to other European nations, the second and the third pillars indicate that it is recognized in France too that there are limits to what the state can do. The imputed mission orientation of the French research institutes, and thereby their lack of market orientation, has probably contributed to the current problems the system faces: its isolation, its difficulties in commercializing results and its human resource problems. The sheltered position of the public research institutes, their close ties to the state and their independence from both the private commercial sector and the universities, has fostered their transformation into ivory towers. The policy discussions in France show that the opinion still dominates, at least among civil servants, that the state should set the course for the economic system, but that the negative side-effects of this demand attention and countervailing action.

Striving for Decentralization

France is a highly centralized country. Not only is economic and political power concentrated in Paris and its surroundings, but also the research community has its most important base there. More than half the researchers working in the private sector and about 40 per cent of those in the public sector are based in Île-de-France, the district around Paris. In the last ten years or so, France has edged towards decentralization, and this is now the main theme in the reorganization of public administration. R&D is one of the policy areas where the regions have been given more power. France's 22 regions are very heterogeneous in their structure and needs, and therefore decentralization seems warranted. In order to keep the regional initiatives consistent with the plans of the national government, the regional governments and the national government conclude a *Contrat de Plan Etat-Région*. These contracts are not only meant to safeguard the consistency of the regional plans with the plans of the national government, but they must also ensure that the national programmes are tailored to specific regional needs. In addition to the regions, several large cities have some autonomy with regard to science and technology policy. The other governmental levels have virtually no involvement in science and technology policy.

Together the regions spend about 1200 million FF a year on innovation incentives, technology transfer and the creation of high-tech enterprises. The decentralization process seems to be having some teething problems. It has increased the need for co-ordination between the (individual) regions and the central government but this has not been satisfactorily solved. One of the civil servants we interviewed complained that he does not know the names, budgets and objectives of the regional civil servants responsible for the matters for which he is responsible on the central level. The regions (at least in R&D matters) do not have unified representation in Paris, and hence, if needed, each region has to be contacted individually. Also, since R&D matters are new to the regions, they are inexperienced in the field and early indications are that they are mainly copying state level policies from 10–15 years ago. The civil servants also said that 'almost daily' they encounter situations where region(s) duplicate what is already done at the state level, leading to unnecessary waste and misallocation of scarce resources. Budgetary restrictions do not leave the regions untouched. The government states that spreading R&D activities across different regions, a government objective for a very long time, will continue only if it does not decrease efficiency.

Reform of Higher Education

What happens when the market starts to govern the system of higher education? You get product differentiation and end up with a two-tier system. On the one hand, there will be facilities for the elite offering top quality education. Selection will be severe and attendance will be expensive; thus entrance will be restricted to the upper classes. Universities serving this segment of the market will charge high fees, but they will recruit top-level teachers from all over the world and will offer excellent conditions. On the other hand, there will be education for the masses. It will be of mediocre quality, more or less adequate for its purpose, structurally underfunded. The universities catering for this part of the market will no longer be able to afford activities that do not pay off in the short term, such as fundamental research and the more costly types of courses, and will not offer their students the newest teaching methods exposing them to the opportunities of ICT. A two-tier system of higher education, once in place, consolidates for generation after generation a stratification of society into social classes.

France has a system of higher education that could naturally develop into such a two-tier structure. The pressures to let the market rule the educational system are mounting and the division between *grandes écoles* and universities, though developed in the past for completely different reasons, already shows many traits of one between education for the élite and for the masses. The Commission Attali, set up by the government in 1997 to develop plans to reform the system of higher education, starts its analysis from confronting these observations with the ideal of an education system that is equally accessible to all students, whatever their background or social class. It puts forward as its main objectives for the future to reinforce the French system of tertiary education by ensuring equal opportunities for access and by reorientating the system towards the needs of the students instead of the needs of the state, which no longer occupies the central position in economic and industrial life.

TOWARDS A EUROPEAN MODEL OF HIGHER EDUCATION

The report *Towards a European Model of Higher Education* has been published by a high-level commission headed by Jacques Attali (former adviser to President Mitterand and former director of the EBRD) and consisting of scientists and company directors. Although the commission did not suggest a fundamental reform of the two-tier system of higher education, it recommends some profound changes.

The present system of higher education, the commission writes, though of high quality, is vulnerable. The universities have had to accommodate the demographic shock of the past decades, stretching their resources to accept ever growing numbers of students, whereas the *grandes écoles* have been able to reinforce their position as centres of excellence. Currently, however, the universities seem ill-prepared to meet the demands of our times. More than a quarter of students leave the university without a degree. Those who leave with a degree often find that their acquired knowledge does not meet the demands of the labour market.

Teaching quality suffers because of lack of recognition for teaching in the reward system. Research quality is not always up to standard either. Reasons are to be found in ineffective administrative procedures, lack of co-ordination and failing evaluation procedures. The *écoles*, having grown much slower than the universities, have increasingly used the opportunities to cherry-pick among their applicants. Their students are primarily the children of high-ranking civil servants and managers of large enterprises. It is now almost impossible for a youngster educated at a deprived high school in a poor suburban area to enter a top *école*.

The system is confronted by a number of challenges. First, the development of ICT and other new technologies are turning around teaching methods and skill requirements alike. Second, higher education is no longer an extension of the state apparatus for the recruitment of high-ranking personnel. The ties between the education system and the state need to be reshaped. Third, the ties between education and private business also need a change. France is relatively weak in those industries that in other countries have developed as spillovers from universities (the ICT industry; biotechnology firms). Fourth, the system has to prepare for the demand for permanent education. Degrees lose their value over time as knowledge progresses, and labour is increasingly mobile between functions.

To meet these challenges, the commission proposes to make the system more transparent, competitive, flexible and responsive, and less socially discriminating. The most important recommendation is to harmonize the system of degrees and bring it more in line with Anglo-Saxon practice: a bachelor's degree (*licence*) after three years, a master's (*nouvelle maîtrise*) after five and a *doctorat* after eight. Universities and *grandes écoles* will use the same system, thus facilitating transfer between the two systems. In higher education more attention is to be paid to research experience and practical 'on the job' training (mainly through apprenticeships). *Grandes écoles* must open up to less-favoured groups, for instance by adopting a quota system of some kind.

Some recommendations pertain to the organization of higher education. To create networks and to optimize the allocation of means, the commission proposes the establishment of Provincial University Poles (PUPs). Excellent university departments

will be assembled into eight regional networks ('poles') to promote co-operation and harmonization. The local administration
of institutions is to be reinforced and the relationship with MENRT
is to be ruled by four-year contracts. Evaluation is to be organized more regularly, systematically and independently, and the
outcomes should translate into consequences for budgets.

To some extent the proposed changes are of a cultural nature.
The system has to become more orientated towards the needs
of individual students by helping them to make choices and to
reach an education level that fits their capacities. Universities
and *grandes écoles* must foster entrepreneurial spirit. Researchers and teachers must become more mobile, both geographically
and functionally.

Public–Private Co-operation and Exploitation of Results

One of the central weaknesses of the French R&D system is its rigid structure. The different parts of the public knowledge infrastructure, the universities,
the *écoles d'ingénieurs* and the research institutes, operate in relative isolation and do not exploit opportunities to bring the fruits of their research to the
market. In the US, the graduates and teachers of one single institution, the
MIT, are founders of about 4,000 enterprises that together provide the liveli-
hood of over 1 million people. These spin-off firms together produce more
wealth than the country ranking twenty-fourth on the list of nations ordered
by their GDP (Attali, 1998). Also in the US, the software industry, which has
become the third-largest in the country, has its roots entirely in the universities. Biogenetics is likely to follow a similar development.

This dynamic interaction between the university system and the private
sector, this commercial exploitation of what started in fundamental research,
is regarded with envy in France. Apparently, the barriers between the French
public knowledge infrastructure and the private sector are so high that they
stem the flow of knowledge before it reaches the market. At the scientific
level, France has a strong position in 66 technologies. It has a leading position in the industrial application of only 24 of those, leaving the rest to others.
In 1994, CNRS received FF20 million in licence fees, which is negligible
compared to the more than FF13 billion state support it receives annually.
Among the 50 largest software firms in the world, only two are French; there
is no French firm among the ten largest producers of computers; there is only
one French enterprise among the first hundred biotechnology firms and one
other among the ten producers of semiconductors.

This isolation of universities and research institutes in the world of academia, separate from the world of commerce, is manifest in the weakness of France in emerging and hi-tech industries. French exports of products from hi-tech industries in 1991 were about half of that of the UK. It is also manifest in the volume of research that private firms fund to be executed by the public knowledge infrastructure. Just over 10 per cent of the funds of public research laboratories come in the form of research contracts, of which only part are contracts from industry. In 1994, contract research for private industry covered 3.1 per cent of the expenditures of the public research institutes and 4.3 per cent of the institutes of higher education (Guillaume, 1998).

So what are these barriers between the public knowledge system and the private sector that stem the knowledge flow and leave it commercially unexploited? A wide array of answers is offered in documents from CIRST and in a report by Henri Guillaume, honorary president of ANVAR, to ministers of education research and technology and of industry and finance:

- A first problem, encountered in other countries too, is the mismatch between the supply of and the demand for knowledge. This is partly an organizational problem. Research organizations are structured on a monodisciplinary basis, whereas firms have needs and problems that touch many disciplines simultaneously. As the need to turn to the market for funding and to acquire a market orientation has been limited so far, the necessary provisions to deal with practical multidisciplinary problems, such as transdisciplinary networks of research organizations, have been weakly developed. The public research system does not have a single 'front door' which can be approached with complex multidisciplinary problems, and through which the knowledge network may be entered. This has led a number of French firms to turn to foreign research organizations, such as the German Fraunhofer Institutes which are better adapted to their needs.
- A second problem is formed by legal barriers that so far have prevented too close relationships between publicly funded researchers and private industry. Currently, these researchers have been forbidden to hold shares or sit on the boards of companies with which they have research links.
- A third problem is the reward system in the knowledge infrastructure. The system primarily rewards academic success as measured by publications. The consequences of this are an increase over the last 12 years in the number of publications and a decrease in the number of patents obtained.
- A fourth factor is the absence of a vision and strategy on the part of the state on the problem of how to guide, co-ordinate and evaluate the public research effort.

- A fifth barrier to the creation of wealth on the basis of research results is excessive bureaucracy and lack of transparency of procedures. For instance, there are more than a hundred publicly financed technology transfer organizations.
- A sixth problem is the excessive concentration of public financial support on a limited number of industrial groups and sectors. In 1994, some ten grand industrial groups related to the military sector received 83 per cent of the public means spent on industrial R&D in the private sector, while they executed only about one-third of total business enterprise R&D. This poses the question whether this type of spending is effective, now that production for military purposes is shrinking and the engine of economic growth and employment is in production for consumer markets.
- A seventh barrier is the lack of venture capital and seed money.
- An eighth, and allegedly the most important, barrier is of a cultural nature: it is the lack of industrial competence and entrepreneurial spirit among French scientists. This argument, by the way, is not exclusive to the French debate; it is heard all over Europe.

MAIN RECOMMENDATIONS OF THE GUILLAUME EVALUATION REPORT

The report of Henri Guillaume, published in 1998, on technological research and innovation, starts out from the observation that: 'Our country commands a first-class scientific and technological potential, but the coupling of its inventions and knowledge with industrial activities happens less easily than in the US and in Japan.' It speaks of the national innovation system as trying to develop some speed with the brakes firmly applied at a time when international competition is accelerating. Guillaume notes that his diagnosis of the situation is not fundamentally different from the one drawn up by the OECD in 1985. He draws attention to the fact that the French state lacks a coherent overall strategy for publicly funded research and development that encompasses the activities of the various research institutions. The universities and large research organizations determine their strategies more or less in isolation, without any mechanism to ensure cohesion in the national R&D effort. This leads to a lack of co-ordination in effort, a lack of transparency of the system and a lack of correspondence between the public R&D effort and private sector demand.

The recommendations of the Guillaume report pertain to four issues. The first point is the development of an overall strategy for science, technology and innovation policy with due attention to the issue of exploitation of research results. The report recommends the establishment of a national Centre for Technological Research charged with the development of research strategies that result in the formation of public–private research consortiums linking firms and public laboratories. This centre should also create a number of competence centres that act as 'front doors' where industry can gain access to the research network. Furthermore, it recommends focusing public spending on industrial R&D on the creation of innovative firms, the promotion of innovation and technology transfer to SMEs, and the reinforcement of links between the public knowledge system and the private sector. Also it suggests an evaluation and reconsideration of the current buoyant spending of public funds on R&D in large firms in the defence industries.

The second issue is the reinforcement of the links between the public knowledge infrastructure and the enterprise sector. Here the report presents a vast array of suggestions, among which are: instruments to stimulate the mobility of researchers and their involvement in private sector initiatives; the rewarding of researchers on the basis of other than academic criteria (for example, mobility, relations with business enterprises, consultancy activities, patenting activity); the reinforcement and streamlining of policy instruments that stimulate and support firms to employ (young) researchers; the creation of interfaces between universities and enterprises and of 'incubator' structures from which to launch new firms.

The third point is the improvement of organizational means for the transfer and diffusion of technologies. The report recommends a number of policy measures to streamline and simplify the instruments of technology transfer and to evaluate the arrangements on a regular basis. It argues for a professionalization of technology transfer centres in universities.

The final issue is the improvement of fiscal and financial instruments to reinforce business enterprise R&D. Here it is proposed to create two national funds to provide seed money for start-ups in information technology and biotechnology. Also the report appeals for a new impulse to venture capital and for an increase in the accessibility of tax deductions for R&D expenditures.

French R&D Policy and the European Union

Despite a generally profound awareness of France's position on R&D matters in comparison with other industrialized nations, the attention of French R&D policy makers is largely directed at the national scene. The main French policy document on science and technology policy (*Projet de Loi de Finances*, 1998) is remarkably brief on EU R&D co-operation. The issue is dealt with in less than a page in this almost 200-page volume, and on this page a policy stance is hardly put forward. France is critical about the evaluation procedures of European programmes, which it considers unsatisfactory; also it wants to see their administration decentralized. Generally speaking, the French attitude towards EU R&D programmes differs little from those in other large EU countries. France realizes that due to its status as a major member state it has the leverage to influence the course of EU R&D policy. As the MENRT is responsible for contacts with the EU, such powers lie there. The civil servants we talked to were, however, of the opinion that France does not use this leverage effectively enough and that more emphasis should be put on influencing the direction of EU R&D.

Recapitulation

The French science and technology policy debate echoes a number of the same themes and arguments as can be heard in other European countries, such as the danger of falling behind the US and Japan in research intensity, the weak performance in emerging technology-based industries, the importance of boosting the technological performance of SMEs. Particularly prominent in the French debate, however, is the emphasis on the rigidity of the public R&D infrastructure, the weakness of private R&D and the dependence of businesses on public provision and stimulation of R&D, and the disparity between scientific potential and the ability to exploit results commercially. We have the impression, though, that France lags behind countries like the UK, the Netherlands and Finland in the process of turning the conclusions of this debate into policy practice. For instance, it is generally acknowledged among French policy makers that SMEs are important as engines of innovativeness and employment growth. However, to our knowledge there has not yet been a turn around of the traditional practice of concentrating R&D support on a handful of large industrial enterprises, some of which are state-owned. Also the process of turning public research institutes into market-orientated organizations working for clients on a contract basis seems to be still at a relatively early stage.

Philosophies that are prominent in a country are not only reflected in the issues that do come up in policy discussions but also in the issues that do not. Philosophies also reveal themselves in the blind spots, through the questions not asked and the arguments not put forward, especially when those are prominent in other countries. The issue of market failure as a guiding principle to determine where the state should intervene hardly surfaces in the French debate. The idea that the main purpose of technology policy is limited to creating conditions conducive to private sector innovation does not seem to fit French concepts of the role of the state. In general there are few arguments heard in favour of a retreating government in R&D matters. For instance, we did not come across plans for large-scale privatization of public research facilities as have been carried through in the Netherlands and the UK. Also, the necessity to devise policies to prevent domestic R&D facilities moving abroad and to attract foreign investment in R&D facilities, a matter with some prominence in German and Dutch policy discussions, does not figure as a policy issue here. The fact that the idea of attracting foreign R&D or of having to worry about keeping existing facilities within the country is foreign to the French policy discussion also reflects the dominant role of the state in R&D matters.

4.4 POLICIES

The French administrative structure is relatively centralized and technology policy is more mission-orientated than in most other European countries. This explains the development in France of a number of policy tools by the government to keep a firm grip on the activities of public research organizations, while at the same time delegating the responsibility for realizing objectives to these organizations and making them accountable for their actions. The main instruments of the state to steer the public knowledge infrastructure are the contracts between the Ministry of Education, Research and Technology and the research organizations and the *Grands Programmes*.

Contracts as Management Tools

The French central government uses three main types of contracts to manage its relationships with others concerning R&D matters: bilateral contracts with public research organizations and with regional governments, and trilateral contracts between the state, universities and CNRS:

1. The *Contrats d'Organismes* are concluded between research organizations and the Minister for Education, Research and Technology. The
government introduced this management tool which establishes the relationship between the various ministries and the public research institutes.
It is a means periodically (every four years) to redefine and clarify the
mission and the strategy of every single research organization. It is also a
means to reconsider the position of institutes in relation to the total of the
public knowledge infrastructure, thereby improving the coherence of the
R&D system as a whole. In the negotiations about the contract, the state
has the opportunity to express its strategic priorities for the different
fields of research and to ensure that its needs are being taken into account. The negotiations on contracts started in 1994, and the first contracts
were concluded in 1995. The various contracts deal primarily with research strategy and the contents of research programmes. Other issues
that are touched upon in the various contracts that have been signed so
far are, among others, the system of research evaluation, development of
overhead costs, human resource management, and mobility of researcher-
ers. Also included in the contracts are the objectives with regard to
collaboration with other research institutes and universities, partnerships
with private companies, and technology transfer activities. The contracts
constitute a general framework for subsequent budget negotiations. They
also provide an important element in the *ex post* evaluation procedures of
research organizations.

2. The *Contrats de Plan Etat-Régions* are concluded between the central
government and the regional governments. The contracts deal with a number
of issues; they also have a research component. The regions have some
responsibility and a limited autonomy with regard to science and technology policy. In order to keep regional initiatives consistent with the objectives
and strategy of the national government, regional governments and the
national government conclude a contract. These contracts are not only
meant to ensure consistency between national and regional programmes,
they also help to tailor national programmes to specific regional needs. In
general, the procedure to arrive at a contract starts with the government
making a list of its main objectives, taking the respective strengths of the
regions into account. This list is sent to the *prefets* (heads of a region) and
to the directors of the main universities and research institutes. At the
regional level plans are drawn up and discussed with all organizations
involved in science and technology. Then the plan is negotiated with
central government authorities and, after approval by both parties, enacted.
Topics in the present contracts are the establishment of *universitées
thématiques*, and the expansion of the number of researchers located in the
regions away from Île de France.

3. The *Contrats Tripartites MENRT–Universités–CNRS* govern the links between the universities and the research groups and facilities that belong to CNRS. These contracts refer, for instance, to the laboratories within universities that operate with research staff and finance from the CNRS budget. They are also used to enhance the coherence of research activities on a national level and to ascertain a strategic orientation in line with the national objectives of science and technology policy. Among the themes currently addressed in contract negotiations are the mobility of researchers, the commercial exploitation of research results, collaboration with other research institutes and an improvement in the quality of theses.

Les Grands Programmes

The French government aims to provide direction to the national R&D effort. One tool comprises the so-called *Grands Programmes Technologiques* (GPTs). These programmes are characterized by an objective defined in terms of the development and construction of 'technologically complex objects', such as aeroplanes, missile launchers, satellites and nuclear power stations, for which the state is, either directly or indirectly, the main client. State involvement is considered of great importance as the mastery of these technologies 'constitutes for France the guarantee of its independence and helps to articulate its economic and political weight in the European context and in the world' (*Projet de Loi de Finances*, 1998). The GPTs cover four fields: space; aviation; nuclear technologies; and information technologies (electronic components, software engineering and telecommunication), and have both civil and defence purposes. They are considered of strategic importance as they must ascertain France's independence in a military sense, with respect to energy supplies, and in access to key technologies. To help France to gain command over these technologies, it is felt that large-scale state support is indispensable. The GPTs are considered an expression of the long-term commitment of the French government to invest in the necessary technological and scientific development, in order to maintain a leading position in these fields of innovation.

By far the largest GPT is the space programme (see Table 4.5); the civil part of it has a value of over 18 per cent of the BCRD, the state budget for civil R&D. The civil expenditures flow about 95 per cent through CNES (including the French contribution to the European Space Agency). The aviation programme has only a relatively small civil component. Together the space and aviation programmes consume almost half of the R&D expenditures of the Ministry of Defence. The nuclear and the ICT programmes account for only 15 per cent and 10 per cent of the defence R&D budget

Table 4.5   *Expenditures on* Grands Programmes Technologiques *in 1994*

GPT	Civil share (%)	Military share (%)	Volume (FF billion)
Space	56.4	43.6	13.3
Aviation	19.1	80.9	9.4
Nuclear energy	42.3	57.7	7.1
Telecommunication	48.0	52.0	5.0
Total	42.2	57.8	34.8
Volume (FF billion)	14.7	20.1	

Source: *OST Indicateurs 1998.*

respectively. Most of the budget for the energy programme goes to CEA, which spends it on collaboration with EDF (producer of electricity), Framatome (builder of power plants) and Cogema (producer of reactors). The telecommunications programme is much more diverse than the others and involves a larger number of contributing organizations.

The big civil and military programmes (GPT) of the French government have been dominant in French national R&D for a long time. The GPTs are considered to be important as triggers for innovation and for co-operation between research institutes and large industrial enterprises. However, some critical notes have been put forward counterbalancing the alleged successes of the GPTs (Duby, 1995). Sectors that are heavily supported by government, for instance aviation, are generally no more competitive (in terms of number of patents and of degree of autosufficiency, that is, the ratio of domestic production and consumption) than sectors which received hardly any support from the government, such as pharmaceuticals. While the programme on nuclear energy made France number one in the world in this field, the French IT sector failed to develop to world class, despite heavy government support. Apparently, the success of the various GPTs differs quite widely. Duby (1995) suggests that a GPT is most likely to be successful if there is a sound knowledge base in a specific field available to start from, and if the industry applying the technologies is not operating in a regular market environment but in a field with a natural involvement of the public sector. Another important weakness of the GPTs is their general failure to involve smaller firms and to diffuse their results to SMEs. SMEs are not much involved in the GPTs; only 4.4 per cent of the total R&D expenditures of SMEs is financed (directly or indirectly) out of the *Grands Programmes*, compared to 18.5 per cent of the R&D spending of large companies.

Support for Private Sector R&D

Next to the GPTs mentioned above there are a number of other programmes, called *Grands programmes industriels*, aimed at meeting specific needs of industries. They are also used to forge links between enterprises and the public knowledge infrastructure. Usually they take the form of a wide-ranging project with an average duration of five years and involve several ministries and state agencies. The programmes started since 1990 show an equal divide between the domains of transports, life sciences and material sciences. In 1997, the four most important programmes concerned: (a) applications of biotechnology to the benefit of health, the environment and the agrifood sector; (b) development of new means of road and rail transport (such as 'hanging' rapid trains, new buses and tramways); (c) new applications of chemical technologies in industrial processes; (d) how to deal with asbestos. The first two programmes have a budget of FF70 million per year, the third FF50 million and the fourth FF10 million.

Besides support programmes directed at the development of technologies, there are also procedures to stimulate the demonstration of new techniques. The instrument *sauts technologiques* (technological jumps) aims at demonstrating the industrial feasibility of innovative and ambitious technologies for a new product or process. The initiative for a project comes from private firms: usually they approach a public research institute to set up a joint project. The subsidy on such demonstration projects can go up to 50 per cent.

The study *100 Key Technologies* that was launched by the Ministry of Economic Affairs, Finance and Industry (MEFI) in 1995 identified 136 different technologies that are considered to be important for the development of the French economy. The study confirmed the strength of French research and its weakness in transforming all that knowledge into successful industrial innovations. Together with ANVAR, the Ministry selected about 50 technologies to be stimulated by public means. An appeal has gone out to industry to come up with proposals for projects that will help to improve the position of French manufacturing industries through the development and use of these technologies. Companies are stimulated to develop strategic alliances with other companies or with research institutes, and to set up innovation projects that will enable them to master these new technologies. The emphasis is on strategic co-operation and network building and the procedure follows a bottom-up approach, which is less common in France. The project is supported financially by MEFI through ANVAR.

Exploitation of Results

It will come as no surprise that French policy documents currently pay a lot of attention to policies targeted at stimulating the commercial exploitation of research results. Every institution of higher education nowadays must have an organization in place, not just for technology transfer, but for the actual industrial and economic exploitation of results. It has to develop a policy to deal with the problem of the commercial utilization of results, taking account of its own characteristics and the needs of its environment. Focal points of this strategy have to be: support for SMEs; development of technology transfer centres; participation in technology diffusion networks; creation of innovative business firms; and temporary or permanent mobility of researchers to the private sector. Agreements on these issues are included in the contracts between MENRT and the institutions and results are subject to evaluation. A recent evaluation of exploitation policies has shown that in the case of most universities a lot remains to be done.

Economic exploitation of the research results of public research institutes usually comes about through partnerships with public or private counterparts. Usually there is some sort of contractual framework. Organizations which belong to EPST have the legal possibility of participating in firms or of concentrating commercial activities in branch establishments. A number of special legal forms have been developed that permit research institutes and universities to co-operate with or take part in enterprises (GIP: *Groupement d'interêt public*, and GIS: *Groupement d'interêt scientifique*). Guillaume (1998) remarks that the legal possibilities of, for example, GIP have hardly been used so far. Also government recognizes that an innovative enterprise developing as a spin-off from research institutes, for instance as a young PhD leaves the institute, is a phenomenon encountered only rarely in France (*Projet de Loi de Finances*, 1998). It is therefore stressed that research organizations should do everything in their power to support commercial innovation projects of their researchers. A possibility would be the development of 'incubators' within institutes that would provide young inventors with information and expertise, advice on legal and commercial issues, and on management, finance and technology.

Fiscal and Financial Policy Instruments

According to CSRT (1995), a lack of (venture) capital is one of the main causes of the high failure rate among French high-tech start-up companies. However, an initiative has been launched to make it attractive to invest in small but promising innovative firms. Recently the *Fonds Commun de Placement pour l'Innovation* (FCPI) have been created. Under this regulation, households (not firms or banks) which buy shares of a FCPI receive significant tax reductions. The FCPI are managed by banks and are invested in a number of small innovative firms. When firms apply for recognition as an innovative firm allowed to participate in the scheme, ANVAR has the task of certifying whether this firm fulfils certain criteria (one of which is that the firm should not already be listed on the stock market).

The Crédit d'Impôt Recherche (CIR) is a generic instrument not directed at specific technology fields or firms. A certain percentage (depending on the region where the firm is located) of the *increase* of corporate expenditures on R&D can be deducted from the corporate tax or income tax bill. Because of its simplicity and speed, CIR is relatively popular among the very small and small companies. The checking of applications occurs later within the context of the usual fiscal checks. Recently the number of applications has been diminishing as R&D budgets have been kept constant or even slashed due to the economic downturn.

4.5   CONCLUSIONS

France's R&D structures and policies are well in line with the popular view of French étatism: the state is prominent in setting directions and its relationships to the large industrial organizations are close. This is evident from the partition of the total R&D effort between the public and the private sector, where business R&D has reached the level of public R&D expenditures only as recently as 1995, and from the prominence of the *Grands Programmes*. It is also reflected in other ways: the reduced emphasis on bottom-up procedures to arrive at innovation, the less-developed formal consultative procedures on policy directions. Also, there seems to be less worrying about position in international competition for the location of private R&D; for instance, there is less attention to creating favourable conditions to attract inward investment in R&D facilities.

France spends a considerable amount of money on R&D; it has a relatively high ratio of R&D expenditure to GDP of 2.34 per cent and in absolute numbers it is the fourth-largest spender in the world after the US, Japan and Germany, but ahead of the UK. Along with the UK and the US it devotes a large share of its budget to defence R&D. The quality of the French science base is considered adequate. The main problems in France concern the rigidity of the structure of the knowledge system. On the one hand, the system is not built to respond in a flexible way to market needs and demands from society; it is built to perform tasks determined by the government. On the other hand, the structures are difficult to manage financially as many of its expenditures, such as wage bills, tend to grow autonomously. The rigidity of the structure makes it difficult for policy makers to find room to manoeuvre and to find the means to invest in renewal of the system, for example, to boost commercial exploitation of the fruits of research and to make optimal use of the technological capacities of the country for wealth creation.

Currently, the largest overall organizational change in French administrative procedures is the increase in regional powers. Decentralization is now considered to be one of the most important challenges to the French government. This has created a situation of flux, as regions take on duties that previously used to be those of the central government and have now become shared responsibilities. There were signs that in the R&D field the increase in regional powers, at least in the early stages, came at the cost of duplication and loss of efficiency.

REFERENCES

Attali, J. et al., 1998, *Pour un modèle européen d'enseignement supérieur*, Paris.

Chesnais, F., 1993, 'The French national system of innovation', in R.R. Nelson (ed.), *National Innovation Systems: A Comparative Analysis*, Oxford/New York: Oxford University Press.

CIRST, 1997, Documents on the MENRT website: www.education.gouv.fr.

CSRT, 1995, *Rapport annuel du Conseil supérieur de la recherche et de la technologie sur l'évaluation de la politique nationale de recherche et de developpement technologique*, Paris: Conseil Supérieur de la Recherche et de la Technologie.

Duby, J.-J., 1995, 'Les Grands Programmes Technologiques: sont-ils utiles à l'économie?', *La Vie des Sciences*, **12** (2), 151–7.

Guillaume, H., 1998, *Rapport de mission sur la technologie et l'innovation*, (www.finances.gouv.fr/innovation/guillame/), Ministère de l'Economie, des Finances et de l'Industrie, Paris.

OST, 1998, *OST Indicateurs 1998*, OST Paris.

Projet de loi de finances pour 1998: État de la recherche et du développement technologique, 1998, Imprimerie Nationale, Paris.

Riel, C.J. van, 1995, *Technologiebeleid in enkele Europese regio's; zeven reisverslagen*, Achtergrondstudie nr.7, AWT, The Hague.

5. Finland

5.1   INTRODUCTION

Finland's remarkable recovery from a deep economic recession which started at the end of the 1980s and its rapid integration into the economic and administrative structures of the European Union after the demise of the Soviet Union has attracted attention to this small country on Europe's north-eastern rim. The recent rise of Finnish Nokia to become one of the top three telecommunications equipment manufacturers in the world is a prime example of the country's economic resilience and technological potential. Finland has undergone its most dramatic recession since independence in the first part of this decade during which the economy contracted by more than 10 per cent. Even in 1996, after three years of export-led economic recovery and growth rates of 5 per cent per annum, unemployment remained high at 18 per cent. The recession had two main causes: the economic and political collapse of the USSR, formerly a major trading partner of the politically neutral Finns; and the vulnerability of the small specialized industrial base of the Finnish economy. There is a wide consensus within the public sector, and more generally in society as a whole, that the industrial structure of Finland is too specialized. Finland stands on a wooden leg: the pulp and paper industry and the related engineering industry dominate manufacturing. The pulp and paper industry's share of exports is over 30 per cent. If the value of pulp and paper industry related engineering products such as paper-making machines is added to this, wood, pulp and paper-related products account for more than 50 per cent of Finnish exports. Today, knowledge-intensive sectors account for an increasing share of Finnish exports.

The consequences of the recession were far reaching. Like its Scandinavian neighbours, Finland has always had an extensive social security system that was successful in achieving the government's social objectives but was considered to be a burden on the country's economic recovery (OECD, 1995). In recent years, a renewed emphasis on individual responsibility and the recognition of limits to the state's abilities to provide a safety net has swept over Finland. Political support for the existing system has been reduced and policies are slowly changing. Nowadays, the general conviction is that Finland's best strategy is to rely on a highly educated workforce and a well-functioning research infrastructure producing the technical progress deemed vital to the competitiveness of Finnish industry. In education, for example, policies are being geared towards flexibility and 'true' competence, as opposed to the traditional emphasis on formal qualifications (seen as a legacy from the era when Finland was part of the tsarist Russian empire).

Finland's geographically isolated position on the northern border of the European continent, neighbouring an economically and politically unstable Russia, contributed to the conviction of the Finnish government that large efforts are needed to keep up with the rest of the world. It helped science and technology to move upward on the policy agenda. Finland, as a small, open economy with relatively scarce resources and high labour costs, has a number of low-cost neighbours nearby. This has provoked large Finnish firms like the paper and pulp manufacturers UPM(-Kymmene) and Metsa-Serla to move production facilities and employment abroad, while decreasing employment in Finland. It is generally agreed that Finland cannot compete with the wage levels or input prices of countries like Estonia, and that thus the only way out is to invest in technology and infrastructure. The overall aim of government is to steer the country towards an information-based society. Great hopes are placed on the development of the electronics and telecommunications industry. Its successes in recent years are seen as evidence that the industrial structure can be changed and that Finland is capable of creating new areas of competitiveness. On the positive side, the changes in the former Eastern Europe and especially Russia open up new possibilities that, however, have only just begun to be realized. The main links so far established have been to service the St Petersburg area (which has a population as large as that of Finland) and to use Estonian firms and labour for subcontracting. The poor state of the Russian (transport) infrastructure has led to an increase in transito-traffic, that is, traffic via Finnish ports to Russia.

Traditionally in Finland, government policies in many fields were heavily influenced by considerations of regional development. Finland is a small country with approximately 5 million inhabitants, very unevenly distributed across the vast territory (which is only slightly smaller than Germany). The urban region around Helsinki is the political, economic and cultural centre of Finland. Although only one-fifth of the Finnish population live in this area, it produces one-third of the country's GDP. This uneven distribution of wealth and population induces a constant stream of industry and people from other regions to migrate towards the Helsinki region, thereby further depopulating the rest of the country. The fear of widespread social disintegration of the countryside has always induced Finnish policy makers to look at the regional consequences of policy measures. Science and technology policy was also evaluated from a regional policy perspective, and this is still reflected in the present national system of innovation. For example, the large number of universities and the many regional offices of national institutes can be considered remnants of the regional dimension of past science and technology policy. These days, things have changed radically. Regional policy and other policies like education and labour market policies are now subject to the overall goal of creating a science and technology based economy. The concept of the 'national innovation system' has moved to centre stage and has been extended to encompass other types of policies that earlier had not been regarded as R&D related.

Finland has a strong tradition of political consensus, which is illustrated by the fact that since the elections in 1995 the government consists of such different parties as social democrats (the largest party in the country), socialists, conservatives, environmentalists and representatives of the Swedish minority party. The consensus orientation of Finnish politics is also illustrated by the well-developed partnership relations between the private and public sectors. Representatives of special interests (most often industry and labour unions) are members of various government councils and committees. An important example from the domain of R&D policy is the Science and Technology Policy Council (STPC). Its members include the Prime Minister, the Ministers of Education, Trade and Industry and Treasury, as well as representatives of universities, labour unions and industry. Government representatives also often participate in committees set up by special interest groups. Given the tradition of political consensus seeking, policy proposals that finally come into effect are usually broadly agreed upon, and changes of government do not have dramatic consequences for the general line of policy. The view that science and technology are crucial to the development of the Finnish economy is currently widely shared and this is not likely to change with a change of government.

In 1995, total R&D expenditures were about FIM12.5 billion, equalling 2.3 per cent of Finnish GDP (see Table 5.1). Since 1985 R&D expenditures

Table 5.1   Distribution of R&D expenditures, 1985–95 (%)

	1985	1989	1991	1995
Enterprises	58.7	61.6	57.0	62.8
Universities	20.9	19.3	22.1	18.3
Other public[a]	20.4	19.1	20.9	18.3
% of GDP	1.58	1.83	2.07	2.30

Note: [a] Mainly government research establishments.

Source: OECD, *Economic Surveys 1994–1995*.

as a share of GDP has steadily increased. This has been partly due to a decrease in GDP at the beginning of this decade. The recession is also reflected in a lower share of private sector R&D in total R&D expenditures in 1991 and 1993. Apparently, the private sector reacted more strongly than the public sector to the recession. However, in 1995 as economic recovery started, private sector R&D expenditures grew at a higher rate than public R&D. The more sluggish reaction of public policy to economic vagaries might be due to budget rigidities or to a deliberate policy to stabilize efforts in the science and technology field.

Private R&D expenditures are distributed unevenly across different sectors. About 87 per cent of total private R&D expenditure is accounted for by manufacturing, of which more than 50 per cent comes from the electronics industry. Currently, most R&D is still carried out by the ten largest industrial firms. Recent history imposed neutrality upon the Finns in matters of global political strife; therefore, Finland's research infrastructure is not burdened with large outlays for defence R&D. Finnish defence R&D (about 2 per cent of total R&D expenditure) is the smallest of all our case study countries.

5.2 STRUCTURES

As in the other countries we have covered, there is a small number of main players who together determine R&D policies and a large number of interests that are involved in the preparation and execution of these policies. The main ministries are those of Education, Trade and Industry and the Treasury. An important, though non-executive, role is played by the Science and Technology Policy Council, the purpose of which is to ensure the coherence of different sectoral policies. However, the Finnish system equals (if not exceeds) the Dutch in co-ordination, involving all major interest groups in the official structures of the R&D process. Both industry and trade unions have representatives on the Science and Technology Policy Council and as a matter of routine on committees established to study particular issues. The involvement of other ministries in the planning of the R&D process is less visible, but they (notably the Ministry of Agriculture) encompass parts of the R&D system in that some government research institutes are situated in these ministries. Figure 5.1 outlines the organization of public R&D in Finland.

Finnish science and technology policy does not originate in isolation but is organized in co-ordination with other policies. Some even tend to view other policy areas, such as employment, trade, education, energy or regional issues, as subordinate to science and technology. In such a situation, where every interest is organized and every organized interest has a say in the matter, and where the issue at hand is considered to be related to almost every other area

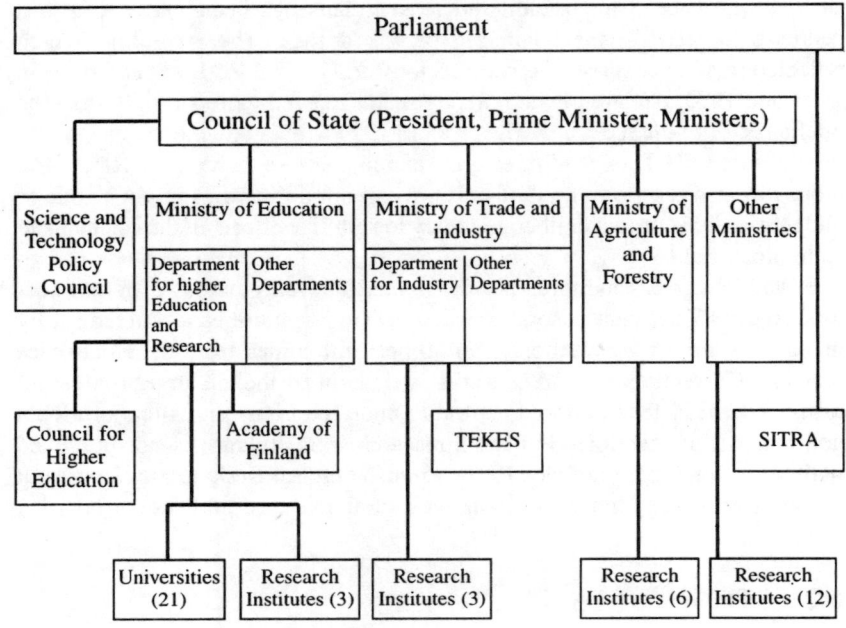

Source: Science and Technology Policy Council, 1993.

Figure 5.1   The structure of public sector R&D in Finland

Table 5.2   R&D expenditures per ministry in 1998

Ministry	Expenditure (FIM million)	% share
Ministry of Education	2784.7	37.9
Ministry of Trade and Industry	2657.0	36.1
Ministry of Social Affairs and Health	676.8	9.2
Ministry of Agriculture and Forestry	445.7	6.1
Ministry of Finance (Treasury)	197.8	2.7
Other ministries	588.4	8.0
Total government	7350.4	100.0

of economic policy, it is in co-ordination that the most substantial problems lie. In general, this involvement creates a broad consensus on the technology issue and a lot of stability in technology policies. The downside of this stability, however, is that in a radically changing environment, as in the early 1990s when the economy contracted by over 10 per cent, the system is slow to respond.

Ministries

The Ministry of Education, Science and Culture commands the largest share of the total government budget on R&D, around 40 per cent. Its responsibility is to plan and monitor education at all levels and to maintain and co-ordinate the public scientific research infrastructure. The Ministry of Education also has responsibility for designating and funding the 'centres of excellence' at universities and research institutes. In general, research institutes and the universities receive part of their funding for both education and research directly from the Ministry. Other research funds of the Ministry of Education are channelled through the Academy of Finland, which is largely autonomous. Universities are independent organizations and they, like the Academy, negotiate their budgets with the Ministry but have considerable freedom in using the resources allocated. A number of research institutes fall under this Ministry. These institutes usually have a mission orientated towards a specific sector or problem area and their work is often contract based. The budget of the Ministry of Education grew at a rate below the average over the 1990s. In the period from 1991 to 1995 the budget of the Ministry of Education actually decreased at a rate of 1.6 per cent annually, while average budget growth over the same period was about 1 per cent. Since 1996 the budget has increased significantly, but still well below average growth.

The second-largest spending department on R&D, with a budget share of around 35 per cent, is the Ministry of Trade and Industry (MTI), which is mainly responsible for applied research and development. In the first half of the 1990s its budget grew at rates above 7 per cent annually, thereby stressing the importance the government attaches to supporting near market research and transfer activities. MTI's main policy vehicles are the Technology Development Centre (TEKES), responsible for planning and financing private R&D, and the Technical Research Centre (VTT), which carries out applied research. Until recently, MTI operated 19 regional offices to finance and advise SMEs on non-technical issues. The technical questions were left to the local TEKES offices (which is also subordinate to MTI). Now, with the establishment of so-called Labour and Industry Centres, desks for non-technical and technical issues are integrated and the government opts for a 'one-stop shop' concept. Furthermore, MTI is responsible for industrial subsidies. Some of these, such as a number of regional subsidies, are inherently geared toward preserving existing economic structures, while others are aimed at supporting new initiatives and structural change. The emphasis over the last few years has been shifting from the former to the latter. A task recently given to MTI is to harmonize all subsidy schemes for businesses that are operated by different ministries.

The task of the Treasury is largely the same as in other countries, although, because of the recent severe recession, it is carried out with ever more vigour: to hold the purse strings. To date, however, this does not seem to have hurt R&D spending too much as government R&D budgets have grown faster, or during the recession years decreased slower, than overall government spending. The overall budget setting process in Finland is called 'frame-budgeting', whereby each ministry is given a frame within which it can decide relatively independently how to allocate funds for various purposes. This system is close in spirit to that operated in the UK. First, the government agrees on a grand total of government spending. Then the budgets of the ministries are decided after discussions within the Cabinet and between the ministries and the Treasury. Finally, the budget proposal of the government is taken to Parliament, which makes the final decision. Individual MPs can make suggestions to change the proposed budget and there is a vote over the final package.

R&D budgets are formally determined relatively autonomously within each ministry. It seems, however, that the Treasury sometimes takes an interest in the factual contents of policies; it was, for example, strongly involved in the discussion on the 'information society' as one of the R&D policy objectives. Over and above this, the Cabinet can make specific commitments that then materialize in the budgets of different ministries (that is, it ties funds within the budget of a ministry). The agreed procedure is sometimes difficult to carry out. For instance, setting the budget of the Ministry of Education is complicated by the fact that this Ministry first must have (internal) budget discussions with individual universities before it knows its own spending total. Frame-budgeting, in the words of some civil servants, facilitates 'vertical integration' (that is, coherence within the budget of a ministry), although it discourages 'horizontal integration' (that is, the attainment of objectives that are not wholly the responsibility of one ministry). The achievement of R&D objectives is a prime example of a shared responsibility and can and occasionally does suffer.

Other ministries are also involved in R&D policy as far as their field of interest is concerned. Most of them have their own research institutes. Every ministry can also contract out research to institutes of other ministries. The Ministry of Agriculture and Forestry has a 6 per cent share in the government R&D budget, mainly to finance research on forestry. Although forestry and forest products are still very important to the Finnish economy, the budget of the Ministry decreased by almost 2 per cent over the period 1996–8.

The Ministry of Foreign Affairs plays an important role in matters of international scientific and technological co-operation, such as the EU or Nordic scientific and technological research programmes.

Science and Technology Policy Council

The Science and Technology Policy Council is a body directly advising the Cabinet and the president on matters of science and technology. The Council's main objective is to guard the overall objectives and policy coherence when different ministries concentrate on implementing their own policies. As stated by the Council itself (1996), its 'object of development is the national system of innovation and its different sectors, with their internal and external interactions'. The Council is a high-level body, its members being the Prime Minister, five ministers, representatives of the Academy of Finland, directors of TEKES and of VTT, and representatives of universities, labour unions and industry. The Council has no executive powers but, due to the political significance of its members and the wide incorporation of different interests, its recommendations carry some weight. It is a good example of the institutionalized way different interest groups are incorporated into the decision making system. If conflicts arise, the aim is to arrive at a solution that all interest groups can accept. The incorporation of special interests guarantees the support of those groups at later stages of policy implementation, but sometimes at the cost of a slow arrival at decisions.

Public expenditure on R&D amounted to FIM5580 million in 1996 (see Table 5.3). This figure rose to FIM7350 million in 1998. Government funds flow mainly through five channels: the Technology Development Centre (TEKES), the universities, the Academy of Finland, the state research institutes and various subsidy schemes.

Table 5.3   Government expenditure on R&D, 1996

	TEKES	Universities	Academy of Finland	State research institutes	Other	Total
FIM (million)	1460	1535	500	1160	910	5580
Shares (%)	26	28	9	21	16	100

Source: Academy of Finland, *Annual Report 1996.*

TEKES

The Technology Development Centre (TEKES) is the main vehicle of Finnish R&D policy. Over a period of ten years its budget more than tripled and by 1996 was almost as large as the R&D budget of the institutes of higher education (see Table 5.3 and Figure 5.2). This rapid growth of the budget

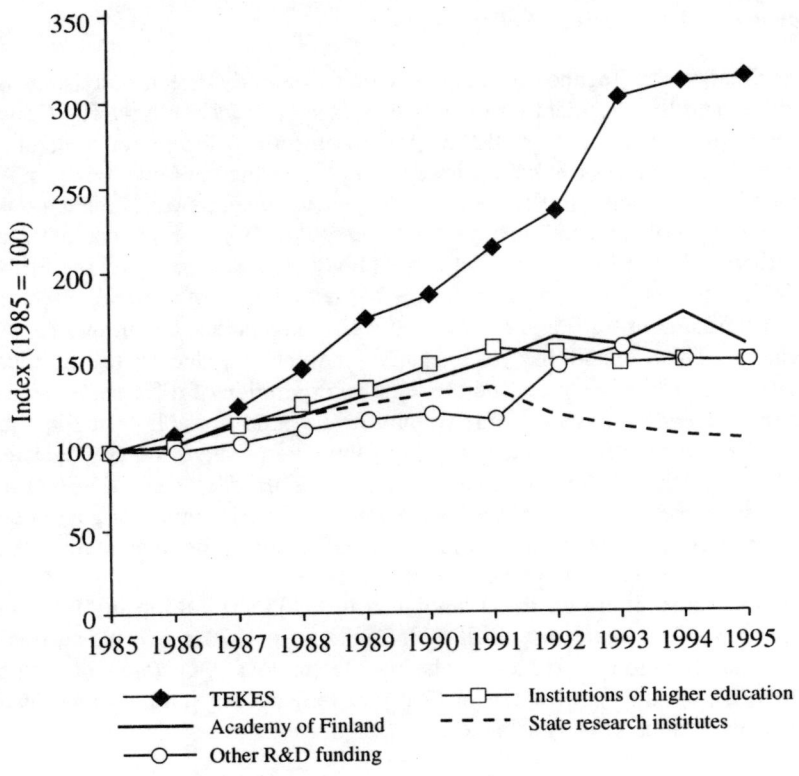

Source: Academy of Finland, *Annual Report 1995*.

*Figure 5.2   Government R&D expenditures through different organizations,
                1985–95*

reveals an increasing interest on the part of the Finnish government to stimulate innovation through the demand side and to subsidize the knowledge
infrastructure by sponsoring the knowledge demand of business firms. There
is some ground for doubt as to whether TEKES's budget will remain at the
current level because the Ministry of Trade and Industry, under which it
operates, faces budget cuts.

Since its foundation in 1982, TEKES has taken a central position in matters of planning and financing technical research and development. TEKES
does not carry out research itself. Its general objective is to improve industry's ability to renew its technology and its technological standard by
supporting companies in technological matters both with financial (soft term

credits and subsidies) and non-financial means (advice and training). TEKES implements the technological programmes of the Finnish government and its international collaboration programmes and hosts the Finnish VALUE Relay Centre and the Finnish Secretariat for EU R&D. TEKES is also strongly involved in the so-called Labour and Industry Centres that are located in regions throughout Finland. To give an impression of its size: TEKES finances about 10 per cent of Finland's total R&D (Numminen, 1996), had an annual budget of FIM1460 million in 1996 and currently employs about 200 people. It spends about half its budget on product development grants, 15 per cent on product development loans and 30 per cent on applied technological research (including space and energy research and international co-operative projects). In 1997, just over 40 per cent was aimed at production and energy technology, 30 per cent at information and communication technology, and slightly less at chemistry and biotechnology.

In addition to TEKES there are other smaller intermediaries such as Technology Centres and Technology Transfer Companies. Technology Centres are established under the responsibility of the Ministry of Trade and Industry in university cities. Their objective is to create new employment by offering cheap locations near universities and research centres for newly established companies. Currently there are nine of them, and 90 per cent of Finnish R&D is carried out in the cities where they are established. Also located in these cities are several Technology Transfer Companies, such as Finntech Oy (Espoo), Oulotech Oy and Tamlink Oy. These companies have expertise in the commercialization of innovations (patenting, licensing, launching of high-technology start-ups) and the organization of research projects involving several partners.

EVALUATION OF TEKES

Zegveld and Guillaume (1995) have carried out an evaluation of TEKES. The result of that evaluation was quite favourable, stressing that TEKES is a widely respected and well-known organization in the Finnish national system of innovation, and that the portfolio of TEKES's activities is, in general, in line with the needs of industrial growth. It is an important instrument in the battle for structural change as it increases the competitiveness of the export sector and helps to create new economic activities.

Notwithstanding the positive tone of the report, the evaluators proposed several changes to TEKES's procedures:

- More attention should be paid to the commercial success of the projects. This clearly implies the development of formulated targets and built-in monitoring and evaluation procedures.
- TEKES should broaden its activities over a larger number of companies (especially SMEs and technological start-ups). This implies that TEKES has to provide the services needed by SMEs.
- In supporting large firms, TEKES should focus more on such aspects as technology transfer to the economy, co-operative developments, the creation of spin-offs and the up-grading of subcontractors.
- The current system of development loans should be converted into a revolving fund with a specified risk profile.
- Together with large firms, TEKES should set up industrial cluster projects in product development involving SMEs and public research institutions.
- TEKES should be more actively involved in the formulation of future international programmes at the EU level.

Currently steps are being taken to implement a number of these recommendations.

Academy of Finland

The second important intermediary in the Finnish national system of innovation is the Academy of Finland. The Academy is the central financing and planning body in basic research. Three-quarters of the research financed by the Academy is carried out at universities, one-fifth at public or private research institutes. Formally the Academy of Finland is under the control of the Minister of Education, Science and Culture, but in general it acts largely independently. As a central government agency the Academy has much influence on the formulation of science policy. It consists of a central board, a central administrative body and four research councils (Culture and Society; Natural Sciences and Engineering; Health; Environment and Natural Resources). The budget of the Academy of Finland slowly increased over the years until 1992, since when it has stabilized more or less, resulting in a budget of FIM500 million in 1996. In the mid-1990s the Academy commissioned a 'science watch' project, providing guidelines for the Academy itself, research institutes, universities and policy makers. There are currently initiatives to arrive at a better co-ordination between the activities of the Academy at the basic end of the spectrum and TEKES at the applied end and at co-operation through joint R&D programmes.

The combined budget of TEKES and the Academy of Finland has increased substantially from the mid-1980s onward and now comprises about one-third of the Finnish public R&D budget. By increasing the budget share of these kind of intermediary organizations, the Finnish government has opted for the creation of a certain distance between itself and the day-to-day running of R&D support programmes, between policy inception and policy implementation. Funding through TEKES and the Academy makes R&D support more flexible and better able to adapt to new demands and developments.

Universities

There are serious worries regarding the efficiency of the Finnish university system. With a population of 5 million, Finland has 21 universities. Only one-third of those have over 6000 students and many do not offer a full curriculum in all main subjects. The reason for the fragmentation of the university system is to be found in past regional policies, when universities were founded to vitalize social and cultural development in remote regions. Policy has changed since then, but efforts to increase the size of units by merger have met with fierce resistance and hence not borne fruit. Of the total budget of Finnish universities of about FIM5000 million in 1996, only some 3 per cent was directly related to performance indicators with regard to research quality, such as peer evaluation, quantitative efficiency indicators and impact measurement.

As everywhere, the issue of knowledge supply meeting the demands of society and of the business community in particular is a matter of debate. It is alleged that 'the effectiveness of the use of the university knowledge base by the industry continues to be of major concern' (VTT, 1995b). The overall aim of the most recent development plan for the universities is to increase the effectiveness of the use of the university knowledge base. Long-run trends are towards more systematic graduate education, tighter focus on specific projects, the promotion of centres of excellence in both teaching and research, and more budgetary flexibility and responsibility. Forced by budget cuts, Finnish universities do compete more often than in the past for outside financing. Among the important outside sources of funds are TEKES and the Academy of Finland, both government funded.

Research Institutes

Research institutes carry out research in specific areas such as forestry, agriculture, the environment, development of the information society, consumer behaviour and public health. The largest research institute in Finland is the Technical Research Centre (VTT). It employs more than 2500 people, had a turnover of almost FIM1000 million in 1995, and is the largest of its kind in the Nordic countries. VTT consists of several independent units. As stated in its mission, VTT is

> an impartial expert organization that produces technical and techno-economic research services in order to improve the competitiveness of companies, to diversify the structure of Finnish industry, and to develop the infrastructure of society. In fulfilling its mission, the primary role of VTT's units is to carry out research and development work, technology transfer and testing. (VTT, *Annual Report*, 1995)

VTT is linked to the Ministry of Trade and Industry, but the private sector is also represented on the board. Other important research institutes are the Institute for Forest Research, the Institute for Agriculture and the National Health Institute. In total there are more than 20 research institutes, some of which are very small.

The government largely finances these institutes. In 1996, total direct funding of research institutes was FIM1164 million, about one-fifth of total government spending on R&D. Currently, reforms are being implemented in the relation between the spending ministries and the research institutes that are subordinate to them. The system of input financing will be transformed into a system of output financing in which ministries will become clients of the research institutes. The latter, in their turn, will have to compete and co-operate with universities and institutes in the private sector as they bid for contracts. Research institutes under the new arrangements will be able to sell their services more easily to clients other than the ministry to which they belong, for example, other ministries and private sector customers.

SITRA

An area of perceived weakness in the Finnish R&D structure is the lack of venture capital. The view of government is that the banking system provides insufficient funds for technology-based projects with above-average risk. The Finnish National Fund for Research and Development (SITRA), controlled directly by Parliament (see Figure 5.1) is a government initiative to remedy the situation. It invests in high-technology companies and grants loans. SITRA's investments vary in size from FIM500,000 to FIM15 million. Usually SITRA buys newly issued shares. Generally, a company will have to leave SITRA's investment portfolio after five to ten years. SITRA does not limit its contribution to financial investment, but it also tries to inject some expertise (financial or sector-specific) by taking a seat on the board of directors. SITRA also supports the transfer of technology from the research community to business through licences and patents. Currently, direct investments are being phased out gradually; they will be replaced by participation in venture capital funds that will be set up at SITRA's initiative together with private equity investors.

5.3 PHILOSOPHIES

An active technology policy in Finland first started to develop in the mid-1960s, when increasing international trade put strong pressure on the Finnish economy. During the following years the budget for science and technology stimulation steadily increased, and several organizations were founded (SITRA in 1967) to stimulate the establishment of new enterprises. VTT was reformed in 1970 in order to make it better orientated towards the needs of industry. Finnish R&D policy entered a new phase when TEKES was established in 1982 and the government started to pay attention to the transfer of technology. At the beginning of the 1990s, when the economy was in deep recession and the budget deficit widened, the government opted for a new strategy. It shifted the emphasis from the support of old structures and industries to the creation of new economic activities, and planned to increase science and technology expenditures to a level equal to 2.7 per cent of GDP in 1997 and 2.9 per cent in 1999. This implied a need to raise both public and private R&D expenditures. Public R&D expenditures are to be limited to 40 per cent of total R&D expenditures. To finance the increase in public R&D expenditures without widening the budget deficit the Finnish government plans to invest the proceeds of selling several state-owned companies.

In Finnish policy making, the use of the concept of the 'national system of innovation' was introduced at the beginning of the 1990s. Currently, the organization of the national system of innovation is directed towards the benefit of the 'individual's well-being and intellectual growth, culture and welfare, and a balanced development of the economy and employment' (STPC, 1996). Finnish science and technology policy puts a strong emphasis on objectives like well-being, intellectual growth and culture, on the social and psychological aspects of technological development. This tendency is also found in Germany and the Netherlands, countries similar to Finland in that corporatism is a prominent feature of the political culture.

Among Finnish policy makers, the 'national system of innovation' is used as a comprehensive term, stressing the entanglement of science, technological progress and social change. It is understood to cover not only the institutions that generate progress in sciences and technical innovations, but also those that account for social and cultural innovations. In practice this means that other fields of policy making that are not usually seen as part of the science and technology area are considered in relationship to scientific and technological development. At its widest, R&D policy is considered to encompass topics such as industrial development, labour market policies, regional development, trade policies, education and environmental policies. In general the government strives for a flexible national innovation system which is able to respond quickly to changing circumstances. Improving the interaction between science and technology policy and other policy areas is one of the significant ways of doing so. Improved interaction opens up new possibilities for innovation and maximizes the social benefits of innovation. An example is the Centres of Expertise programme which aims to renew regional economic structures by stimulating technological progress. Preferably the interaction between different policy areas takes place within industrial clusters. This has proved a viable means of developing co-operation and interaction, as an integrated approach to telecommunications (including technical, social and economic effects) has shown.

Scientific and technological development, by becoming a central policy focus, has to some extent replaced regional development as the integrative element in policy making. From the new vantage point, regional subsidies are often perceived as aimed at preserving old economic and social structures, frequently against the tide of history. Support for innovation, even if targeted at regional development, gives priority to dynamism, to the change of economic and social structures in line with the needs of tomorrow. Further changes in the structure of government are currently being considered to strengthen this comprehensive approach to decision making. The tasks of the Cabinet Economic Policy Committee may be extended to include the development of the innovation system (STPC, 1996). As mentioned, the frame-budgeting system, whereby every ministry is allotted its own budget, is not conducive to 'horizontal', interdepartmental co-ordination. There is a lack of co-ordinated planning, with the STPC being the only governmental body taking an interest in the overall structure of public R&D, although without executive power. Though the lack of executive power is partly counterbalanced by the high-level membership of the Council, this does not always ensure a satisfactory degree of coherence in the national science and technology effort.

Today, in Finland official justifications for science and technology policy cite market failure arguments as often as in other countries. Although Finnish policy makers seem to see themselves in a more proactive role in society than many of their colleagues in other countries, the principal aims of Finnish R&D policy are no different from those in other countries: to link demand for knowledge to supply; to let the market mechanism do its work as much as possible; and to compensate if the market mechanism does not yield an optimal outcome.

The overall policy objective is to create a sound knowledge infrastructure to reduce dependence on traditional sectors of economic activity and to transform the Finnish economy into a knowledge-based economic system. In the opinion of the Science and Technology Policy Council (1993), the creation of an adequate knowledge base and a well-functioning national system of innovation is more important to Finland than it is to other developed countries because Finland has to compensate for its geographical isolation. To this end, the government has formulated four main areas of interest (Ministry of Trade and Industry, 1996; STPC, 1996):

1. *Selective public support for R&D* Market failures, particularly the lack of venture capital on the Finnish financial market, have reduced private R&D efforts to a suboptimal level. Therefore the Finnish government supports private R&D, supporting especially high-potential sectors and technologies which have the most profound effects on competitiveness and society. High-potential sectors are considered to be both traditionally strong sectors (such as paper and other forest products) and newly developing sectors, for example, communication and multimedia.

2. *R&D co-operation* Another policy focus of the Finnish government is extension and improvement of R&D co-operation, both domestic and international. Co-operative R&D efforts are deemed vital to Finnish competitiveness. The efficiency of domestic co-operation between the public sector, universities, research institutes, intermediaries and the private sector is to be improved. International co-operation is to be extended to enable Finnish companies and research institutes to anticipate new technological developments in an early stage and to reduce the costs of research. Finland started to participate in international and EU research programmes and initiatives such as the Framework Programmes and CERN at a later date, partly as a consequence of its political neutrality. However, once it had become a member of the EU, integration into these institutions proceeded swiftly, in the expectation that future economic growth would provide the means to cover the extra strain on the government budget without the need to cut domestic R&D spending. In the current economic situation, the increased spending on official international co-operation tends to result in increased financial pressure in other fields of R&D policy.

3. *Education* With regard to the structure of the education system, the government made two important decisions: to establish graduate schools; and to create a new layer of polytechnics. The graduate schools are considered to be an additional and necessary element in the Finnish national system of innovation (Zegveld and Guillaume, 1995). Finland has the smallest proportion of PhDs in science and technology of all Nordic countries, and this is generally perceived as a weakness. The

polytechnics were introduced to fill the gap between medium-level vocational education and universities. In total, 30 schools of this type will be established. The first were set up four years ago. The establishment of polytechnics was controversial: the division of labour between the universities and the new polytechnics is not entirely clear and opinions have been voiced that their creation leads to duplication of existing systems where the number of higher education institutions (universities) is already too large. In the UK this 'binary divide' has recently been abolished; but in Germany and the Netherlands, upper-level vocational education is regarded as very important. As for the content of educational programmes, there is a tension between industry's demand for highly specialized employees and the need for 'flexible' workers. Schemes are worked out that must provide opportunities for life-long education. Government considers vocational training to be partly a responsibility of companies and partly of employees themselves.

4. *The information society*   Like the Netherlands, Finland aspires to become one of the leading information societies in the world. The Science and Technology Policy Council (1996) states that 'an information society is one in which the information and media industries are important business sectors, in which everyone has access to information services and the skills to use them, and in which the procedures and structures of business life and the public sector have been developed with the help of information technology'. This description not only focuses on the technical aspects of the information society but also on the use of information and knowledge. Consequently, the government aims 'to provide every Finnish citizen with basic knowledge and skills needed to understand and use information and communication technologies as a tool in learning, research, work and leisure time pursuits' (STPC, 1996). The National Committee for Information Society Issues was set up to intensify co-operation and information exchange between public and private sectors.

As in many countries, the development of evaluation procedures of past achievements as well as 'forward look' exercises to set the agenda for the future are becoming increasingly important. STPC (1996) notes that these should become regular instruments to help anticipate new developments and inform policy. Almost every policy measure and every organization involved in science and technology policy has been subject to an evaluation study. Often evaluations are carried out by foreign nationals or institutes (frequently in co-operation with Finnish experts or institutes) to guarantee a sufficient degree of independence. In general, recommendations coming out of these studies are taken very seriously. In line with this, in universities there is a lively debate about quantitative measures for the quality and quantity of research (and teaching) produced and, despite some resistance and criticism from within the scientific community, the outcomes of these seem destined to play an ever-increasing role in resource allocation.

5.4 POLICIES

Finnish science and technology policy is mainly diffusion-orientated rather than mission-orientated. It is concerned with the improvement of the knowledge base of firms and the innovative capacities of business firms rather than with large national programmes or projects. This is reflected in the distribution of public R&D funds over socio-economic objectives (see Table 5.4). Almost half of the budget is spent on the general advancement of science. Two other objectives to which substantial amounts of funds flow are the promotion of industry and social policy and services. The largest increase took place in energy R&D. Defence R&D is hardly relevant; it is the only area in which R&D expenditures have decreased since 1996.

Table 5.4   Government R&D expenditures by socio-economic objective in 1998

	Expenditures (FIM million)	% share	% growth (1996–8)
Agriculture, forestry and fisheries	458.2	6.2	14.8
Promotion of industry	1988.8	27.1	25.1
Energy	402.0	5.5	117.4
Defence	101.6	1.4	−11.0
Soil, water and atmosphere	93.0	1.3	42.6
Social policy and services	1334.9	18.2	43.5
General advancement of knowledge	2782.4	37.9	30.5
Space	189.7	2.6	14.4
Total	7350.4	100.0	31.7

Source: *VTT* (unpublished).

Over the last few years means were shifted from the support of old industries towards supporting R&D activities resulting in new industrial activities. This fitted in with the government's objective to eliminate subsidies which distorted competition. Also, attempts were made to make the financial support system more transparent and to eliminate overlap. Consequently, the

Ministry of Trade and Industry reduced the number of subsidy schemes from thirteen to three (MTI, 1995). The Ministry of Trade and Industry not only increased R&D support, but also channelled an increasing share of R&D support to SMEs in order to enhance their innovative capacity. The Ministry also intends to shift financial aid from direct subsidies towards indirect aid instruments such as guarantees, loans and stimulation of venture capital.

FINANCIAL AID BY THE MINISTRY OF TRADE AND INDUSTRY

Tax credits
Currently, there are no tax credits to stimulate R&D. In the past, however, such instruments did exist, but they were considered inappropriate and therefore abolished. There is a preference for instruments to stimulate R&D that are more narrowly focused. However, in their evaluation of TEKES, Zegveld and Guillaume (1995) conclude that there is a need for a system of tax credits in Finland.

Subsidies
Subsidies can be given to technology-based activities, especially those of SMEs. There is no maximum size of the subsidized project. In order to receive a subsidy a project has to have a certain risk. TEKES awards a subsidy if a project is 'far away from the market', although it is not clearly defined what this means. In other cases it might provide a loan.

Loans
Loans are provided to technology-based activities, including technology investments that are 'near the market'. If a project is not successful, the loan may be converted (partly) into a grant. The interest rate is below market rate, while the total loan period is limited to ten years.

Capital loans
Capital loans enable the company that invests in R&D to improve or maintain its initial level of solvency. Interest payments and capital repayments have to take place only if sufficient capital is left. This measure is new, and the Finnish government has set aside FIM70 million to provide capital loans.

R&D Programmes at TEKES

Finland's R&D policy consists of a large set of programmes, each consisting of many small and large projects around one specific theme. There are over 60 different programmes, of which those under the responsibility of TEKES can be classified into two categories: (a) programmes directed at the development of specific products, in which a single company can take part; and (b) public technology programmes for the general advancement of knowledge in a sector or in the whole of manufacturing industry: these programmes are open to groups of companies (within a sector or across different sectors), research institutes and universities. However, TEKES also supports many projects that do not fit into any specific programme, thus making the total support system more flexible.

Public technology programmes come in three types. There are national technology programmes that form

> an important instrument in the precompetitive phase for supporting, establishing and promoting structural change and competitiveness in industry. The programmes are aimed to build the know-how basis of the technology needed for further industrial R&D. National technology programmes often link industrial sectors via cross-sector technology, synergy and diffusion, and increase co-operation between industry, research institutes and universities. The programmes have worked for the implementation of the national technology policy and are also serving as a basis for international technology co-operation. Examples are software technology, biodegradable polymers, machine vision, pharmaceutical technology and mechanical wood technology. (Zegveld and Guillaume, 1995)

Furthermore, there are programmes aimed at specific sectors, in which firms invest in precompetitive R&D in order to raise the general technological level of an industry. Finally, there are cluster programmes, designed

> for industry groupings with a rather limited number of companies. Some of these programmes are largely on SME participation. A limited number of companies are able to focus the programme on key targets based on their technology strategies. SMEs are encouraged to make joint use of existing technologies and often work together with large firms. (Ibid.)

New programmes can be proposed by TEKES itself, and also by companies, groups of companies, universities or research institutes. In this way the system accommodates bottom-up initiatives and puts the responsibility to come up with ideas for research themes that are deemed important for society partly in the hands of companies and research institutes. In response to a proposal, a committee is established that evaluates the feasibility of the proposed programme. If the proposal is approved, an external project leader and a TEKES internal co-ordinator are appointed.

TEKES-Clinics Programme

In 1992 TEKES started up a number of 'clinics' in important technological fields. The clinics operate in the form of 'projects' that are carried out over a specified period of time and are linked to research institutes and universities. They typically have a small budget of around FIM1.2 million and are run by a co-ordinator who maintains an extensive network among both SMEs and public research organizations. The objective of these clinics is to promote the transfer of technology and knowledge in a particular area from universities and research institutes to SMEs. The clinic can help a company with the testing of a new technique, with the practical application of research results, and with getting entry to the services of research organizations. In these cases, it pays half of the costs involved. TEKES started these clinics because it was felt that SMEs struggle with particular problems in innovation. For SMEs, ready access to external knowledge and expertise is often difficult and the complications and the duration of procedures to apply for R&D subsidies is an important barrier. New clinics are established whenever a clear need is expressed; in 1996, 11 specialized clinics were in operation across the country (Korkala, 1995).

An independent evaluation by Autio and Wicksteed (1998) was quite positive about the effects of the clinics programme. The typical SME applicant is not a high-tech company: most applicants are looking for support in decision-making concerning investments in new technology. The clinics perform their intermediary function in a rapid, flexible and non-bureaucratic way. In their approach they are supplementary to other technology programmes.

Centres of Expertise Programme

The Ministry of the Interior of Finland initiated the Centres of Expertise Programme in 1994. The objective of this programme is:

> to facilitate the prerequisites for the location and development of internationally competitive enterprises which require a high degree of expertise. The programme supports regional specialization and assignment of tasks to appropriate centres of expertise. The aim is to enhance the development of the knowledge base by promoting collaboration between new technology based firms and the higher education institutes, research centres and the government sectors authorities. (VTT, 1996b)

The centres of expertise are not institutes in the material sense (with their own buildings and staffs), but virtual organizations carried out as projects under the auspices of the maakuntas.

MAAKUNTAS

After R&D had been made the main guideline for policy decisions, the government concluded that regional policy efforts too should be directed at intangible investments (education, R&D, technology transfer) and the development of new infrastructure. To achieve these objectives at a regional level, the Finnish government created the so-called *maakuntas*, a government layer positioned between local communities and provinces.

One of the tasks of the *maakuntas* (although not their main task) is the development of a regional technology policy. This policy consists of a regional development programme comprising a number of concrete projects (Korkala, 1995). The *maakuntas* were established as a means to create 'one front door' for companies at which all necessary information would be obtainable and through which all applications for support would reach the authorities. Some *maakuntas* have delegated their R&D tasks to companies that were established for this purpose or to technology centres. For example, the company Culminatum Oy in Uusimaa (close to Helsinki) is responsible for the 'centres of expertise' on biotechniques, medical techniques, mechanical wood processing in construction, telecommunications, industrial design and design management, and industrial automation.

Venture Capital

According to Ingman (1995), the Finnish capital markets are still young and weakly developed compared to those in other OECD countries. However, since the 1980s the volume of private venture capital has been steadily rising. In the period 1990–5 the total volume tripled. Nevertheless, Ingman (1995) notes two important problems: (a) in general SMEs are not aware of all the new financial instruments available to them; and (b) the Finnish taxation system traditionally encourages companies to raise loans instead of increasing the company's equity. Therefore Finnish companies have a relatively high debt-to-equity ratio, and a more vulnerable financial structure compared to that of many competitors in the EU countries.

The objective of the government is to strengthen the conditions for venture capital investment by stimulating the supply of finance and by the development of risk-sharing instruments, of expertise on the evaluation of investment projects and of markets for financial derivatives. New types of financial instruments must be experimented with and introduced; necessary legal arrangements are to be developed.

In general, four categories of capital investment funds can be distinguished: bank-related, private, national (SITRA, Kera Oy, Matkailunkehitys Nordia, Suomen Teollisuussijoitus) and regional funds. A heavy state involvement in the provision of venture capital apparently triggered the private capital market. The amount of funds raised increased from FIM250 million in 1994 to FIM300 million in 1995. The government share in the provision of venture capital decreased rapidly. In 1994 the government was still the largest source of venture capital, but in 1995 insurance companies (48 per cent), banks (22 per cent) and pension funds (15 per cent) became more important funding institutions than the government. Important changes were scheduled for 1997, when Finnish law began to allow the issue of preferential shares, preferential capital securities and option rights (European Venture Capital Association, *Annual Report 1995*). On the initiative of SITRA, the Finnish Venture Capital Association was established. This association tries to create awareness among private investors of the investment possibilities in SMEs. It created a new agency service for investment enquiries in order to channel investments by private citizens into the SME sector (SITRA, 1995).

5.5   POLICIES TOWARDS EUROPE

Finland has been a member of the European Union since 1995. Before that there had been a ten-year period of gradually growing participation in several EU (and other international) science and technology programmes: Framework programmes (on a project basis since 1987, fully since the EEA agreement in 1994); EUREKA (since 1985), ESA (associate member since 1987, full member since 1995) and CERN (since 1991). However, Finland only received full rights in EU decision making processes when it became a member of the EU.

International co-operation is mentioned as a priority in almost every government publication on science and technology. EU R&D funds are considered to be a major source of R&D spending even though Finland has not been able (officially) to affect the contents of the first four Framework programmes. EU money is viewed as strictly additional to domestic financing. The Finnish contribution to EU R&D is part of the overall membership payment and is not attributed to any single ministry. In general, Finnish officials who were interviewed are satisfied with the success of Finnish bidders to EU research programmes. However, they also notice a misperception in Finland: EU R&D funding is often considered 'outside' money rather than something the allocation, co-ordination and targeting of which could and should be influenced. Probably the short period of membership accounts for this attitude.

The Finnish government wants EU R&D funding to become an integrated part of the national system of innovation. The aim is for EU research and domestic research to become fully complementary. For a small member state like Finland, EU R&D offers opportunities for which no resources would otherwise be available. Therefore, Finland definitely 'supports efforts to reinforce the status of research in the European Union' (STPC, 1996). The Science and Technology Policy Council states that 'Finland is responsible for its own participation in EU research and for exploiting the results of this research at home' (STPC, 1996). Furthermore, the Council explicitly states that Finland is responsible 'for developing EU research to ensure that it serves as fully as possible the interests of all Europe as well as Finland' (STPC, 1996). As to the organization of EU funding: 'Finland has organized its EU research on the basis of a decentralized model in which responsibility for research co-operation rests with the same body responsible for research in that field within Finland. The whole entity is represented by the EU R&D secretariat and the Research and Technology Section of the Committee for EU Affairs' (STPC, 1996). The Council considers this decentralized model a good base for harmonizing domestic and EU research. The Council also stresses the importance of evaluating the impact of EU funding on Finland's own research efforts.

With regard to the spending of the EU budget in general, the Council argues that the funding of R&D should take a more prominent place. Although the emphasis must stay on industrial and technological research to improve the competitiveness of European industry, more attention has to be paid to humanities and social sciences (especially the utilization of research results). However, an increase in funding should not be distributed across a large number of issues but should be limited to those in which the European Union has competence. The Council appeals for an increasing degree of integration of basic research, applied research and product development, and therefore the need for the EU to fund some basic research. Rather than direct involvement, the Union should stimulate co-operation between basic research institutes in the member states. Furthermore, the funding of the Union's own basic research organizations must be revised. They must be compelled to compete for research funds. As to decision making, the Council stresses the responsibility of the member states and the research institutes to co-ordinate their research efforts. The Union itself should be made more flexible with regard to decisions on research (that is, qualified majority voting on these issues; leaving sufficient uncommitted funds).

The Finnish government is convinced of the importance of international co-operation. However, it also stresses that member states have their own responsibilities, and emphasizes strongly the principle of subsidiarity. In general, the opinions of the Science and Technology Policy Council on EU science and technology policy follow the same lines as those on Finnish science and technology policy. In short, research institutes should compete for funds; more emphasis should be put on humanities and social sciences; total research funding should be increased; research co-operation with non-EU countries should be promoted; and regional funds should be directed towards creating new knowledge-based industrial structures.

5.6   CONCLUSIONS

In Finland, the area of science and technology is considered as a central theme in policy making. Several institutional arrangements, most notably the existence of the Science and Technology Policy Council and the fact that several of the most important ministers are its members, explicitly reflect the importance attached to science and technology policies. It is also reflected in the government's perception of the 'national innovation system' as a comprehensive system of generation and use of knowledge, and, in (at least some) civil servants' views, that science and technology policy should encompass the more traditional policies such as industrial, employment and trade policy. It is not entirely clear, however, to what extent the latter is reflected in practical policy decision making. As in the Netherlands, the word pragmatism aptly describes the Finnish approach towards science and technology policy.

A wide consensus exists that the way to improve the well-being of the members of Finnish society is to invest more in education and the knowledge infrastructure and to lower thresholds and hindrances to the movement of people and information. Not only the importance of the generation of knowledge is stressed, but also its diffusion and utilization. Therefore, the transfer of knowledge to citizens and business firms, especially SMEs, receives ample attention. Compared to the 1970s, when there was a tendency towards more centralization in all fields of policy, the direction has now reversed towards more decentralization (subsidiarity). The view that decision making powers and responsibility for obtaining finance should be delegated downwards within the hierarchy is winning ground. An example of this is changes in the procedures for financing research within government research institutes and universities.

It is generally recognized that the state's ability to assist industry is limited and that policy is best confined to those areas where the state has an explicit responsibility or a comparative advantage: providing education; general infrastructure; and generic remedies against instances of market failure, for example, by stimulating the supply of venture capital when the market fails to do so by using the proceeds of the privatization of state enterprises.

The designers of science and technology policy in Finland, a small open economy, are acutely aware of the dependence of the Finnish economy and its technological infrastructure on international trends and influences. Finnish officials prefer to have an internationally comparable set of policy tools available for the implementation of science and technology policy. Although international trends are not automatically followed, they are monitored carefully and, when needed, own policies are adapted to them. As an example, the number of tools used to intervene in industry is slowly declining, following international trends. Another example is the decline of traditional industrial policy tools in favour of R&D-based tools. On the higher education side, especially with regard to postgraduate education and research, structures and methods of financing are moving towards the Anglo-American model. On the technology side, there is a clear trend towards internationalization and a greater emphasis on company-based R&D. The central role of the Technology Development Centre (TEKES) in the national innovation system illustrates this trend. As to European science and technology policy, the Finnish government stresses the same points as it does with regard to its own science and technology policy: flexibility, market orientation, management by objectives, the use of regional funds for technological progress.

The culture of the Finnish political decision making system is corporatist in spirit. Interest groups have guaranteed access to the system in the form of having members on various committees. The benefits of this are well known in a country like Finland where everybody knows everybody; but the disadvantages were painfully revealed during the recession of the early 1990s when the system altered its course only slowly. Discussions about how to change the system without losing its positive attributes is ongoing but has not yet reached a conclusion. However, the emphasis on continuous evaluation of science and technology policy measures can be looked at as a solution to this corporatist dilemma. Thorough evaluations clarify the discussion, enable policy-makers to keep in touch with the real needs of society, and make possible an adequate reaction on policy failures or changing circumstances.

REFERENCES

Academy of Finland, 1995, *Annual Report 1995*, Helsinki: Academy of Finland.
Academy of Finland, 1996, *Annual Report 1996*, Helsinki: Academy of Finland.
Agricultural Research Centre of Finland, 1995, *Annual Report*, Jokioinen: Agricultural Research Centre of Finland.
Autio, E. and B. Wicksteed, 1998, *Technology Clinic Initiative: Evaluation Report*, Helsinki: TEKES.

Council of State, 1995, *Developing a Finnish Information Society*, Helsinki: Council of State.
Geological Survey of Finland, 1995, *Annual Report 1995*, Espoo: Geological Survey of Finland.
Ingman, T., 1995, *Financing of Young Technology Oriented Companies and Their Growth: Challenges and Recommendations*, Working Paper 6/1995, Helsinki: Ministry of Trade and Industry, Industry Department.
Korkala, P., 1995, *Technologiebeleid in drie Europese regio's: Een internationale vergelijking tussen Aken, Uudenmaan Liito en Gelderland*, The Hague: Ministerie van Economische Zaken.
Ministry of Finance, 1995, *Finland's Way to the Information Society*, Helsinki: Ministry of Finance.
Ministry of Trade and Industry (MTI), 1995, *Annual Report*, Helsinki: Ministry of Trade and Industry.
Ministry of Trade and Industry (MTI), 1996, *A New Outlook on Industrial Policies*, Publication no. 4, Helsinki: Ministry of Trade and Industry.
Niskanen, P., 1995, 'Tutkimus- ja kehittamismaararahat valtion talousarviossa vuonna 1995', *Suomen Akatemian julkaisuja* 2/95.
Numminen, S., 1996, *Evaluation of TEKES Funding for Industrial R&D*, paper prepared for the Six Countries Programme, Gent.
OECD, 1995, *Economic Surveys*, Paris: OECD.
Science and Technology Policy Council (STPC), 1993, *Towards an Innovative Society: A Development Strategy for Finland*, Helsinki: Science and Technology Policy Council of Finland.
Science and Technology Policy Council (STPC), 1996, *Finland: A Knowledge-based Society*, Helsinki: Science and Technology Policy Council of Finland.
SITRA, 1995, *Annual Report*, Helsinki: SITRA.
TEKES, 1995, *Views on Finnish Technology*, Helsinki: Technology Development Centre.
VTT, 1995a, *Annual Report 1995 (Technical Research Centre of Finland)*, Espoo: VTT.
VTT, 1995b, *VTT Group for Technology Studies: Annual Report 1995*, Espoo: VTT Group for Technology Studies.
VTT, 1996a, *Annual Report 1996 (Technical Research Centre of Finland)*, Espoo: VTT.
VTT, 1996b, *Innovation Policy Developments in Finland*, Espoo: VTT Group for Technology Studies.
Vuori, S. and P. Yla-Anttila, 1992, *Mastering Technology Diffusion: The Finnish Experience*, Helsinki: ETLA.
Zegveld, W. and H. Guillaume, 1995, *The Technology Development Centre of Finland, TEKES: An International Evaluation*, Publication no. 5, Helsinki: Ministry of Trade and Industry

6. Germany

6.1 INTRODUCTION

In 1990, after the collapse of the communist system in Eastern Europe, the German Democratic Republic (GDR) became part of the Federal Republic of Germany (FRG). The unification confronted the German government with several problems that were, and still are, difficult to deal with. By the time of the unification, the economy of what then became the *neue Länder* (new states) was practically bankrupt after years of underinvestment and mismanagement under communism, and had to be rebuilt almost from scratch. Not only production facilities and the physical infrastructure needed to be renewed or replaced: the knowledge infrastructure also was not in good shape. Public research institutes and universities suffered from a number of serious problems. They employed too many people, had outdated facilities and were inefficient because of the centralized management of the communist government. Research activities were often aimed at duplicating advances made in the West and were mainly directed at the creation of a more or less autarchic economy. Despite these problems, some institutes and universities reached a high scientific level in some areas. After unification, unemployment in the former GDR increased, leading to social problems and migration to the former FRG. Environmental problems also made a major overhaul of the production system necessary. In order to deal with the various problems, the federal government decided to direct a large, continuous flow of money towards the *neue Länder*.

Over recent decades, industries that were traditionally strong in Germany, some of which had had a competitive advantage over foreign competitors since the beginning of the twentieth century (for example, chemicals, machinery, equipment and vehicles) lost ground against competition from East Asian countries. On top of that, in the years after unification, the whole German economy was hit by recession. The huge costs of unification led to high interest rates and supplementary taxes. German industry, already confronted with high labour costs and a strong D-mark, suffered even more from this additional burden. It cut back employment in order to increase flexibility and reduce costs. Unemployment started to grow at unprecedented rates. At the same time labour unions continued to press for wage increases, even in the former GDR where productivity was still low. The OECD points out that the German economy suffers from social and economic rigidities (OECD, *Economic Survey 1996*). Despite the massive funds directed towards the *neue Länder*, the economy in those areas is recovering only slowly. The slowness of the recovery in the east led to a sense of frustration among the citizens of the former GDR. After 16 years in office *Bundeskanzler* (Prime Minister) Helmuth Kohl was denied another term in September 1998. Elections brought a left-wing government to power which had more sympathy for interventionist economic policies. Whether this recent change in government will affect R&D policy is as yet difficult to predict.

The national system of innovation in Germany, as the third-largest economy in the world, differs from that of the small case study countries such as Finland and the Netherlands. Due to its size and wealth, Germany has for a long time been able to maintain research institutes and budgets for research that were large enough to cover the whole area of basic and strategic research in depth. There was less need to make profound and far-reaching choices as to where to spend money, less need to determine priorities as small countries had to do. Consequently, the German national system of innovation is very extensive. However, in other respects the German system resembles those of the Netherlands and Finland. Although Germany's national system of innovation is comparable in size to those of France and the United Kingdom, German science and technology policy seems less mission orientated and more in touch with the needs of small and medium enterprises (SMEs), and policy mechanisms revolve more around consensus seeking. Compared to France and the United Kingdom, Germany spends little (in relative terms) on defence research.

Germany is a high-wage country that tries to sustain a comparative advantage in international competition by fostering the quality of its workforce, by investment in the productivity of its industrial sector, and by maintaining an institutional and political framework which works well. The downside of German industry being good at what it does is that, at least according to public perception, it is reluctant to take up new innovations, to venture into new markets and products. In its economic survey on Germany (1996) the OECD confirms that the German economy has 'a sound R&D base and is effective in adopting new technologies in existing industries'. However, 'there is concern that new opportunities are not effectively used. One manifestation of this is the deficiency as regards the establishment of firms in new knowledge intensive sectors.'

6.2 STRUCTURES

Germany spends a relatively large share of GDP on R&D, more than the Netherlands and the UK, roughly the same as France and Finland. However, the German ratio of R&D expenditures to GDP has decreased over the last ten years from almost 2.9 per cent in 1987 to 2.3 per cent in 1997. Since the beginning of the 1990s just over 60 per cent of R&D has been financed by the private sector. There has been a shift in the composition of the public share of R&D expenditures: while the federal government cut its R&D expenditures, the state governments increased theirs (Table 6.1). The governments of the 16 *Bundesländer* are becoming increasingly important factors in science and technology policy.

Table 6.1   Partition of R&D expenditures (%)

Funding sector	1991	1994	1997	Growth	
				1991–4	1994–7
Federal government	22	21	19	–4	0
State governments	16	18	18	15	7
Total public R&D expenditures	38	39	38	4	3
Private sector	62	61	62	3	8
Total R&D expenditures (DM million)	76.580	78.909	83.709	3	6
As share of GDP	2.66	2.37	2.32		

Source: BMBF (1998a).

Over the last five years, about 18 per cent of total domestic R&D expenditures has gone to research activities at universities. A somewhat smaller amount, around 15 per cent, is spent on research at public and private non-profit research institutions. This mainly concerns public research organizations like the Max Planck Gesellschaft, the Helmholtz Institutes and the Fraunhofer Institutes. The private sector carries out roughly 67 per cent of total domestic R&D activities. The private and the public knowledge infrastructure are fairly separate circuits. Of the R&D carried out in the private sector, only between 8 and 9 per cent is financed by the state; the private sector finances no more than 8 per cent of the research performed in universities and 3 per cent of that performed in research institutes.

Germany is known for its tradition of corporatism; consensus decision making is one of the prominent features of the German policy making. However,
there are less formal links between interest groups and government departments than in other consensus-orientated countries like the Netherlands and
Finland. Instead, advantage is taken of the high degree of self-organization of
the German economy. For example, industry branch organizations are much
better developed in Germany than in the UK. Interest groups are tightly organized and often have somebody in charge of science and technology related
matters. Sometimes there are detailed procedures, such as strict rules on job
rotation to prevent inbreeding. These people are consulted on a regular but
mostly informal basis by civil servants.

Policy Makers at the Federal Level

Science and technology is a policy area that overlaps many others; therefore
every ministry has some dealings with this subject (see Figure 6.1). However,
the main player in the field is the Ministry of Education and Research
(*Bundesministerium für Bildung, Wissenschaft, Forschung und Technologie*:
BMBF). Next to this ministry, which has a co-ordinating role, there are the
Ministry of Economic Affairs and Technology (*Bundesministerium für
Wirtschaft und Technologie*: BMWi), the Treasury and the Ministry of Defence as important policy makers. The role of the Treasury is similar to that in
most countries: frame-budgets are agreed between spending departments and
the Treasury, and within the constraints of the frame departments are relatively autonomous. This is especially true with regard to R&D activities,
where the Treasury does not formally participate in spending decisions within
the frame at all. When deemed necessary, officials of the Treasury communi-
cate their opinions on particular R&D policies informally. However, the
Treasury minister does try to ensure that spending departments do not launch
schemes that will be difficult to finance later.

In Germany, parliament (the *Bundestag*) seems to play a more active role
in the determination of science and technology policy than in the other case
study countries and seems to have more influence on the details of spending
decisions. There are two important parliamentary committees that have an
influence on federal science and technology policy. The first, the Science and
Technology Committee, provides a discussion forum for MPs and civil servants. The second, the Budgetary Committee, discusses the details of the
departmental budget with every ministry. It is important when it comes to
actual decisions. The Committee has an influence, not only on the grand total
of ministry-level spending (the frame), but also on its contents. It is not
uncommon for the Committee to order a department to shift resources from
one programme to another. One feature of the German budgeting system is

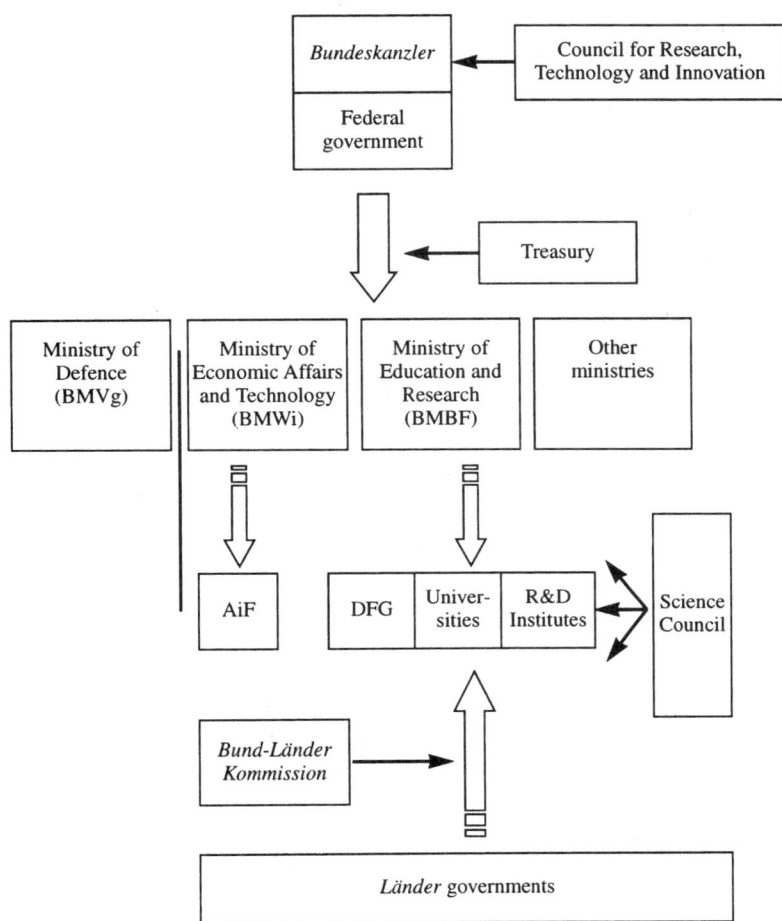

Figure 6.1   *The structure of the German public system of science and technology policy*

that budgets are very detailed and, once set (that is, decided on by government and parliament), they are very rigid. Hence it is very difficult, if not impossible, to shift resources between items within a budget.

Table 6.2 gives an overview of the R&D expenditures of the federal government. The general need to cut government expenditures to fulfil the European Monetary Union (EMU) criteria has lead to a decrease of federal expenditures on R&D over the last two years. The only exception is the R&D budget of the Ministry of Agriculture and Food, which grew by 10 per cent.

Table 6.2   R&D budgets of federal ministries in 1998

	Expenditures (DM million)	Share of government R&D expenditures (%)	Growth rate 1996–8 (in nominal terms, %)
Min. of Education, Science, Research and Technology	10 769	66.3	–1.0
Min. of Defence	2 781	17.1	–4.6
Min. of Economic Affairs and Technology	863	5.3	–14.4
Min. of Agriculture and Food	450	2.8	9.9
Other ministries	1 368	8.4	–5.5
Total federal government	16 231	100	–2.6

Source: BMBF (1998a).

The budgets of all other ministries, especially those of Economic Affairs and Defence, have decreased.

The Ministry of Education and Research (BMBF) has the main responsibility for science and technology policy. Until 1998, it spent almost two-thirds of total federal expenditures on R&D. BMBF was created in 1994, after a merger between the Ministry of Education and Science and the Ministry of Research and Technology. The two departments were united in response to the fact that education and science on the one hand and research and technology on the other are becoming increasingly interconnected, and that an integrated approach to policy issues was needed. In 1998, at the takeover of power by the new coalition, a number of themes in the area of technology policy were transferred to the Ministry of Economic Affairs (BMWi). The areas that were transferred include amongst others: innovation and technology support for SMEs (indirect support only); support for the establishment of new enterprises; nuclear safety; applied energy research; aeronautics research; multimedia for the business sector; and media law. At the same time closer co-operation between the two ministries was announced, the stated purpose being the need to reinforce the links between technological progress and job creation. Most of BMBF's budget is spent on research institutes, universities and the *Deutsche Forschungs Gemeinschaft* (DFG).

Less than one-fifth of the R&D budget of BMBF has been used so far to support private industrial R&D; most of these responsibilities are now with BMWi. Policy measures include credits for SMEs to adopt new technologies, financial support for high-tech start-up companies and for the employment of extra R&D personnel, and the maintenance of centres for information and advice to the benefit of private companies. Because of the particular division of responsibilities, until 1998 the BMWi accounted for only 6 per cent of the federal expenditures on R&D. The activities of BMWi were complementary to those of BMBF and this induced some inefficiency and problems of co-ordination. For instance, BMWi supported the development of civil aircraft, but BMBF too had a large programme on aviation and funded *Deutsche Forschungsanstalt für Luft- und Raumfahrt* (DLR), a large aviation research centre. Both ministries supported innovation and near market research and were concerned with technology transfer, the difference being that BMWi concentrated on SMEs. In 1995, 56 per cent of the R&D budget of BMWi was spent to the benefit of SMEs. The main channel through which BMWi has until now spent its funds is the *Arbeitsgemeinschaft Industrieller Forschungsverein* (AiF).

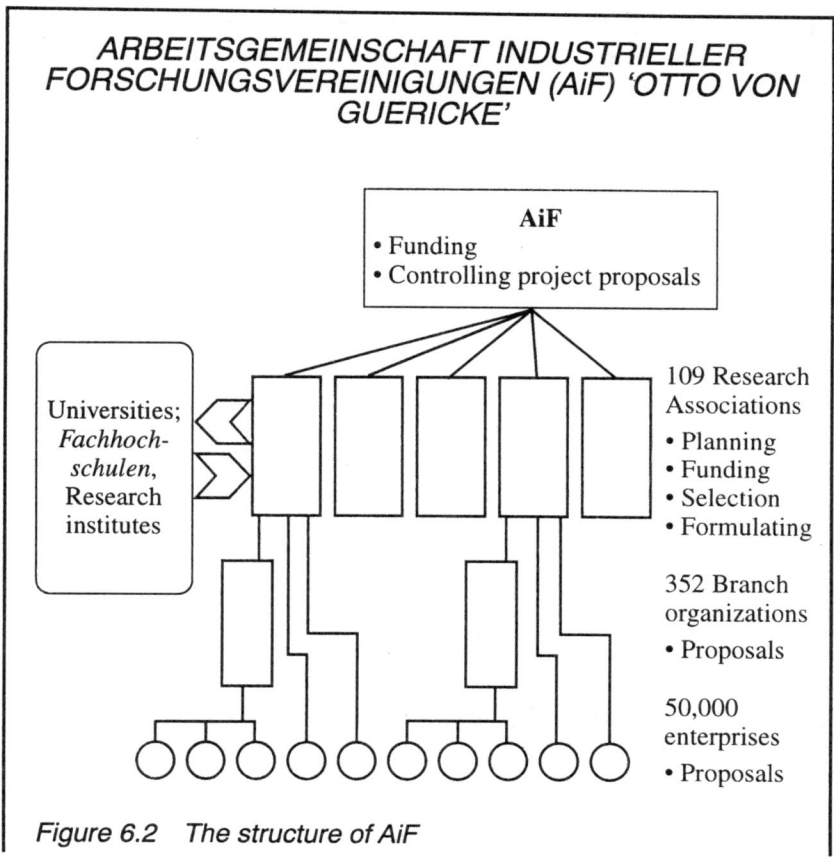

Figure 6.2   The structure of AiF

The AiF consists of 109 different associations of companies or sector organizations all of which stimulate industrial research. It supports applied research projects that are specifically concerned with the needs of SMEs. In general the subjects of these projects are sector specific. Individual companies can propose research projects, but the results should raise the general technological level of the sector and should not favour an individual enterprise. The 109 associations that belong to AiF can either support the research themselves or ask for support from AiF. The Ministry of Economic Affairs finances the activities of AiF. Figure 6.2 gives an impression of the rather complicated structure of AiF.

Other ministries are responsible for R&D activities in their own policy area. Their budgets are mostly rather limited with the exception of that of the Ministry of Defence. The R&D efforts of this ministry account for around 17 per cent of federal expenditures on R&D but these have shown quite some variation over time since 1990, moving between 22 per cent and 16 per cent. In relative terms, Germany spends little on defence R&D (around 0.1 per cent of GDP), but in absolute terms this still amounts to a considerable volume. German military R&D is about 50 per cent of that of the UK and 40 per cent of its equivalent in France.

Universities and *Fachhochschulen*

The German system of higher education is divided into universities (sometimes called *Hochschulen*) on the one hand and polytechnics (*Fachhochschulen*) on the other hand. The universities offer academic and the polytechnics higher vocational education. Universities have about three times as many students as polytechnics. Vocational education combines a sound theoretical knowledge base with a fair amount of practical experience through trainee- and apprenticeships. The integration of training on the job in particular is characteristic of the German system. Polytechnics teach both technical and administrative subjects. The German polytechnics are at the top level of an extensive system of vocational education that has proved very successful in the past in supplying German industry with a highly skilled labour force. Next to its success in interweaving theory and practice, its ease of access is considered a strong point. At the lowest level there are hardly any preconditions to entry. This is cited as one of the reasons for low youth unemployment in Germany.

Despite its merits the system is now under threat. First, due its popularity amongst young people the number of applicants for a training post is larger by far than the number of on-the-job training facilities. Second, it used to prepare students for a broad range of traditional professions, varying from bank employee or welder to engineer. Although the content of most courses is regularly updated, the system as a whole is fairly conservative. It accommodates only slowly to demands for the preparation of courses for new professions. Currently the government is trying to increase the speed of adoption of new professions by the system: it now also offers courses leading up to new professions such as film and video editor and advertising designer. Polytechnics are mainly educational institutions, but it is expected that they will engage more in research in the future (Bauer and Zabel, 1994).

DEUTSCHE FORSCHUNGS GEMEINSCHAFT (DFG)

DFG (the German Research Society) provides financial support to research projects at universities and research institutes. To obtain funds, researchers have to follow a tender procedure. DFG stimulates the training of researchers and collaboration among researchers. In its function, DFG is more or less comparable to the Dutch NWO and the Academy of Finland. Members of the DFG are mainly (but not all) universities and research institutes. The emphasis is on natural sciences, but social sciences are certainly not neglected. Currently, BMBF has plans to transfer some funds from the Blue List research institutes (see below) to DFG in order to bring more competition into the German research system. Furthermore, DFG is being encouraged to become more involved with the transformation of basic research results into applications (BMBF Press Release, 11 July 1996). The federal government finances 50 per cent of the DFG budget. The other half comes from the state governments.

It is not only the system of vocational education which is under strain at present; the facilities for academic education are overstretched as well. The number of students is too large for the present size of the universities. Crowded lecture rooms and a shortage of teachers frustrate both students and personnel. Problems have persisted for years now; the universities' efforts to solve them have affected the quality of research. Because of the heavy workload and the outdated facilities, many researchers are known to prefer to work at one of the large German research institutes or even to move abroad. The number of students in German universities is large for a number of reasons. First, academic education is easily accessible; this is considered an important social achievement and attempts to curtail the right of access meet with strong resistance. Second, in the German system of university education there is little personal coaching. Students are left to their own resources; they do not receive a lot of support and advice from mentors or tutors during their studies, and there is little supervision of their progress. The failure rate is relatively high, and those who do not fail take a long time to complete their courses. The average German student spends seven years at university before graduating (compared to five years for students at the polytechnics), which is about twice as long as the time required to get a first degree in countries following the Anglo-Saxon system.

Research Organizations

After the Second World War, the federal government (together with the state governments) founded a number of research organizations, each of which contains a variety of different institutes. There are five groups of research institutes: the *Helmholtz-Gemeinschaft*, the *Max-Planck-Gesellschaft*, the *Fraunhofer-Gesellschaft*, the Blue List Institutes and the federal research institutes. To give an impression of their size, the total expenditures of these groups of institutes are listed in Table 6.3. As we can see, the research organizations differ in the degree to which they are dependent for basic funding on the federal government. They also receive basic funding from the

Table 6.3   *Total expenditures (DM million, including non-R&D expenditures) and employment of the main German research institute groups*

	Federal: state basic funding	Total expenditure, 1998 (est.)	Federal basic subsidy	Personnel 1997 (fte)
Helmholtz Institutes	90:10	4195	2600	21 611
Max Planck Institutes[a]	50:50	1887	801	11 175
Fraunhofer Institutes	90:10	1350	502	6 620
Blue List Institutes	50:50	1740	674	11 736
Federal Institutes	100:0	—[b]	3485	20 989

Notes:
[a] Except the Institute for Plasma Physics, which is counted as a Helmholtz Centre.
[b] Total expenditure for 1998 is not yet available, but on average is only slightly above federal funding.

Source: BMBF (1998a).

state governments (except the federal research institutes). Besides these basic subsidies, the institutes receive other funds from the federal and state governments as well, mainly for carrying out contract research.

The *Helmholtz-Gemeinschaft* comprises 16 large-scale research institutes (of which three are located in the former GDR). The institutes are largely independent; the *Helmholtz-Gemeinschaft* is the overall organization that links the institutes in a federation. The institutes are funded by the federal government (90 per cent) and the host state (10 per cent); they concentrate mainly on long-term basic research, often involving complex and expensive large-scale research facilities. Most Helmholtz Institutes employ between 400 and 1500 people and have an average budget between DM75 and 300 million (basic funding plus other means). Three institutes are much larger than average: the *Deutsche Forschungsanstalt für Luft- und Raumfahrt* (DLR) is involved in research on aeronautics and space, and merged in 1997 with the German space agency (DARA); *Forschungszentrum Jülich* is active in various kinds of research (life sciences, energy, the environment, IT, materials); *Forschungszentrum Karlsruhe GmbH Technik und Umwelt* deals with environmental technologies, energy, and microsystems technology. These institutes employ more than 3000 people and have a budget of DM625–765 million.

Many Helmholtz Institutes have opened transfer and information offices for industry, and support scientists who seek co-operation with industry. Recent changes limit the independence of the individual institutes and increase the power of the board of the federation of Helmholtz Institutes. Five per cent of the Helmholtz institutional funding will be transferred from the budgets of individual institutes to a general budget under the control of the board. The board will set priorities and recommend areas of interest for all institutes. The institutes can then apply for funding from this budget. Applications will be granted on condition that projects are concerned with industrial or international co-operation, strategic research or the employment and education of young researchers. The government stimulates the institutes to act in a more entrepreneurial way and to build core competencies that should be evident from the budgets they negotiate with BMBF. However, the government will no longer provide detailed guidance on the use of funds. Instead, the institutes and the government will agree on the general objectives of research (BMBF, 1998a).

The *Max-Planck-Gesellschaft zur Förderung der Wissenschaften* (MPG) comprises 71 (largely independent) research institutes spread across the country and a few *Arbeitsgruppen*, mainly at universities in the former GDR. Sixteen institutes are located in the *neue Länder*, most of which are still in a (re)development phase. The MPG institutes perform basic research, mainly in the natural sciences (about 80 per cent of the budget), but also in the social sciences. Most prominent are biological and physics research (one-third and almost one-quarter of the budget respectively). A main part of their mission is to concentrate on research in fields where new methods are required or where universities cannot readily step in (such as interdisciplinary research). To commercialize the results of research MPG has founded the *Garching Innovation GmbH*.

The *Fraunhofer-Gesellschaft* comprises 43 research institutes, of which nine are located in the former GDR. Their main task is to carry out applied research for industry. Although much larger, Fraunhofer is comparable to Dutch TNO or Finnish VTT. German governments (federal and state) tend to turn to the Fraunhofer institutes if they deem it necessary to develop specific technologies that are important for the competitive position of German industry. Currently, Fraunhofer is also involved in the establishment of innovation centres. It runs the *Patentstelle für die deutsche Forschung* which offers help to every institute or individual wishing to apply for a patent. The German government finances 30 per cent of Fraunhofer's budget (of which 90 per cent comes from the federal government) and about 70 per cent is raised on the market for contract research. In the *neue Länder* the government currently still pays for more than 60 per cent of the costs. Local industry is not yet able to carry the full bill for research.

The *Einrichtungen der Blauen Liste* (Blue List Institutes) is a very diverse group of 83 independent organizations. Some carry out research (either basic or applied), while others (libraries and museums) are concerned with documentation and preservation. The institutes of the Blue List are considered to be of more than regional importance, therefore costs are shared equally by the states and the federal government (in most cases). Currently the BL institutes are under threat of losing part of their budget to the DFG. However, it is supposed that they can regain these funds by applying for projects at the DFG. Over the coming years the Science Council will evaluate all Blue List Institutes. The federal government threatens to withdraw its financial support from those institutes that do not get a positive evaluation.

Federal institutes have limited independence from the ministry that finances them. In many cases they are concerned with testing, standardization, promoting German culture and the like. Research and development is often no more than a secondary task; it sometimes accounts for only 10 per cent of their expenditures. R&D activities of federal institutes are generally in support of the main activity of the institute or policy decision making of the funding ministry. The federal government plans to change the way the federal institutes work R&D activities will be carried out on a project base, forcing institutes to become more alert to market developments.

Intermediary Organizations

Next to the extensive research system there exists a large number of organizations that are concerned with the transfer of knowledge and technology from research institutes to the private sector. Most research institutes have founded transfer offices themselves. However, many intermediary organizations operate independently from research institutes and are financed by the state governments. Examples are the Steinbeiss Foundation in Baden-Württemberg and ZENIT and AGIT in (parts of) Northrhine-Westfalia. The *Rationalisierungskuratorium der Deutschen Wirtschaft e.V.* (RKW) operates in all parts of Germany and was established to stimulate the improvement of production processes (*Rationalisierung*). In the former GDR the federal government supported the establishment of 21 technology transfer agencies (ATI) and 13 sector- or technology-specific transfer centres (*Technologiespezifischen und branchenorientierte Transfer Zentern*: TTZ). All these organizations offer help to companies (especially SMEs), try to bring them into contact with research institutes and universities, offer management assistance or access to technical knowledge, and organize training courses for employees or entrepreneurs. They may differ in the way they approach companies or in the sector or technology they focus on, but essentially they are very much the same. It might be wondered whether the number of organizations is not confusingly large. On the other hand, having a large number of intermediary organizations stimulates competition amongst them and perhaps increases the quality of the services they offer.

Advisory Bodies

The *Rat für Technologie, Forschung und Innovation* (Council for Technology, Research and Innovation) reports directly to the *Bundeskanzler* on matters of science and technology. It is a committee composed of scientists of high reputation, captains of industry, chairpersons of trade unions and ministers. Its brief is 'to gain comprehensive insight into important fields of innovation and to make corresponding recommendations. Its focus is on intensifying the dialogue between industry, science, and the state on fundamental questions and aspects or research, technology and innovation.' Its first output, *Informationsgesellschaft: Chancen, Innovationen und Herausforderungen* [The Information Society: Opportunities, Innovations and Challenges], published in 1995, is a report on the opportunities and threats of the increasing importance of information technologies. This publication has been one of the inputs of *Info 2000: Deutschlands Weg in die Informationsgesellschaft* in which the government formulates its goals and initiatives with regard to the future 'information society'. A second report, published in 1997, is entitled

Biotechnology, Genetic Engineering and Economic Innovation: Making Responsible Use of Existing Opportunities; a third report, on *Kompetenz im globalen Wettbewerb* [Competence in Global Competition] was released in 1998.

The *Wissenschaftsrat* (Science Council), established by the federal government and the states in 1957, is the main advisory body on science policy issues in Germany. Its task is to prepare recommendations concerning the thematic and structural development of higher education, science and research. At the request of the federal and state governments it also evaluates individual institutes and advises on proposals for capital investments by universities and polytechnics. After reunification, the Council played a crucial role in drawing up detailed proposals for the restructuring of the research base in eastern Germany. Currently it is engaged in an evaluation of the Blue List institutes. Often the Science Council is concerned with very detailed questions. Its members are eminent scientists, representatives of federal and state governments, and others from public life. The Council has no decision-making power and its recommendations require a two-thirds majority.

Bundesländer

No discussion of German science and technology policy is complete which does not pay due attention to the activities of the states. The states enjoy significant autonomy in R&D matters. In some cases they not only have an interest in R&D issues as represenatives of the public cause, they are also involved as shareholders in large private companies. For example, the government of Lower Saxony is a significant shareholder in the Volkswagen company. Most states have their own policies on education (especially universities and *Fachhochschulen*), research institutes and the transfer of information and technologies to the private sector. Occasionally (but not very often), state policies come into conflict with federal policies. This happens more frequently if a state is governed by parties that on the federal level are in the opposition. Having a different political party in power at the state and at the federal level occurs fairly frequently in Germany as voters like to use their votes in state elections (which are never held on the same day as federal elections) in order to voice their grudges against the policies of the federal government. However, apart from differences in political colour, opposing interests between state and federation seem to account for conflicts as well.

All states have a particular ministry that is responsible for science and technology policy. There is no general rule as to which ministry is primarily responsible in a state. In some states the Ministry of Education is concerned with both science and technology policy, while in others the Ministry of Economic Affairs has the main responsibility for the technology policy part.

Regarding education and most research activities, the states are bound by federal law; but regarding intermediary organizations for knowledge transfer to the private sector, each state may develop a structure of its own. In addition, state governments develop their own research and technology programmes and have their own research institutes. The *Bund-Länder Kommission für Bildungsplanung und Forschungsförderung* (BLK) co-ordinates the efforts of the federal and state governments with regard to education and research. This applies in particular to information on the principles and procedures relating to research funding as well as to plans for new or existing research institutions and projects. The Commission is a forum for discussion on matters relating to education and research promotion. One of its tasks is to develop medium-term plans; it makes recommendations on the budgets of those institutions that are jointly governed by the *Bund* and the *Länder*.

Institutes of education are mainly financed by the state governments. The federal government, however, is responsible for the legal framework (the *Hochschulrahmengesetz*) that applies to all institutes of higher education. It is also responsible for providing students with scholarships if they need financial support. To change laws on higher education and scholarships the federal government needs the support of the senate, which consists of representatives of the state governments. Together with the state governments, the federal government funds large-scale research facilities and investments in real estate and develops specific programmes, for example to increase the number of female researchers or to attract foreign students. Finally, the federal government finances R&D projects in which universities can participate. Federal and state governments depend heavily upon each other with regard to higher education, which can make decision making very complicated.

Science and Technology in the *Neue Länder*

When the GDR joined the Federal Republic of Germany in 1990 the public knowledge system was in bad shape. Research organizations were overstaffed and underequipped, and were inefficient because of their centralized management structure. Management posts were often allocated on the basis of political criteria rather than scientific merit. Research took place in relative isolation from what happened elsewhere in the world. As in other areas of life, it was then decided to extend the system that existed in the western states to the new states in the east. Within a few years the *Max-Plank-Gesellschaft* and the *Fraunhofer-Gesellschaft* established a number of research establishments in the former GDR, sometimes in collaboration with local universities. Despite the adverse conditions, GDR science was of a high quality in certain niche areas. Attempts have been made to preserve these strengths and to build the new knowledge system as much as possible on the basis of the old.

The plans for restructuring the public knowledge infrastructure in the east were developed by the *Wissenchaftsrat*. They evaluated the capacities of the GDR research groups and the demand for fields of technology, and advised on the future domain, size and structure of the research organizations to be created. Regarding the recruitment of personnel, it was decided to have open competitions for all management posts, for which also researchers of the former organizations could apply. For research posts the rule was to strive for continuity, so most vacancies were only open to internal applicants. Nevertheless, an objective of personnel management is to have around 10 per cent of researchers from the west (German or foreign) in order to arrive at a balanced composition of the labour force.

The biggest problem in the former GDR, however, is R&D in the private sector. Research and development proved particularly vulnerable when after unification the economy in the east entered a process of transformation. After privatization, many companies suddenly had to find their way in a market environment, and a new role for R&D was often not immediately obvious, leaving R&D departments in companies hanging in mid-air with no strategy. Also, in the process of privatization many research departments were separated from production departments when the latter were sold off on their own. Employment in industrial R&D decreased from 75 000 to 13 000 people in the first three years after unification. In the German east, 86 per cent of R&D personnel in companies works for firms smaller than 500 employees and half of them for firms with fewer than 20 employees; in the west it is the other way round. Massive state intervention managed to secure a considerable chunk of industrial R&D for the future, but company R&D is still weak. A number of former research departments of enterprises now function independently, accounting for 30 per cent of employment in industrial R&D. A major weakness at the moment is the absence of strong R&D in large companies. Large firms are usually more export orientated than the average firm. They are able to carry out large R&D projects with longer time horizons which draw on fundamental research. They are an essential link between the more academic research of the public knowledge suppliers and the short-term development-orientated research which makes up most R&D activities of SMEs.

It is felt, however, that given the present industrial structure the innovative capacity in the east must be developed on the basis of the potential of SMEs. Therefore current policies to foster industrial R&D are directed mainly at them. There are policy measures to increase R&D personnel in firms and to support R&D projects. There are calls for policies aimed at attracting outside investment in technology-intensive production activities and for policies to strengthen the basis of the independent industrial research organizations that were formerly part of firms. A number of them seem to have developed a capacity to survive in the market.

6.3 PHILOSOPHIES

On the issue of technology policy, the federal government under the Christian Democrats believed its main responsibility was to create an environment conducive to innovation. To perform R&D is in the first place the responsibility of the private sector. There are good reasons why government should support the R&D efforts of the private sector – the market failure argument is as often cited here as it is elsewhere – but the objective of government intervention is primarily 'to help companies to help themselves'. It is as yet difficult to say whether philosophies under the new 'red–green' coalition government of Social Democrats and Environmentalists will be very different, but in general they tend to opt for a more active role for the state. The traditional line of policy is to strive for an integrated and broad innovation-stimulating policy that does not limit itself to the creation of research facilities, but also stimulates all the relevant actors to contribute to national competitiveness. However, the German government is comparatively reluctant to get involved in direct support of private sector R&D (as the figures cited above on government sponsored private R&D show); it is only willing to step in if the private sector is not able to react fast enough or on the right scale to new developments. In this case government policy is aimed at reducing risks, stimulating co-operative research and improving information flows. This attitude fits in well with the concept of a *soziale Marktwirtschaft* (social market economy), a mix of socialist and liberal ideas which in some form or other has been central to German economic policy for the last 50 years (and which currently seems to be gaining in popularity across Europe under the name 'the third way').

Standort Deutschland

A central issue in discussions on science and technology policy in Germany nowadays is the attractiveness of *Standort Deutschland*, Germany as a location for companies to establish production and R&D facilities. In this respect Germany is not unique; governments of all case study countries appear to be concerned about the attractiveness to increasingly foot-loose transnational corporations. However, the issue seems to come up with more persistence in the German debate than it does, for instance, in France. The vehemence of the present German discussion may have various reasons. First, it is probably related to the less mission-orientated policy stance of the German government: the focus is traditionally not on steering the economy in a preferred direction but on competing on the basis of offering better conditions to private firms. Second, it can be related to the fact that the German economy is now in recession and that its traditional industrial strongholds have come under fierce threat from foreign competitors. Third, the need to rapidly create new economic activities from scratch in eastern Germany after unification makes the issue of attracting private sector investment in high-quality production activities all the more urgent. The discussion on the German *Standort* directly touches on issues of science and technology policy. For many firms, the availability of a high-quality supply of knowledge resources, as revealed by the quality of the knowledge infrastructure and of the level of skills and education of the workforce, is one of the main determinants of their decision on a location. In Germany, as in other developed countries, the availability of knowledge resources has to compensate for disadvantages such as high labour costs.

Multinational companies praise Germany for its high-quality research facilities, its highly qualified personnel, its engineering competence and its demand for high-quality solutions that simultaneously satisfy economic, environmental and societal constraints. However, the German *Standort* is also considered to have some significant disadvantages, such as very strict and extensive regulations, especially with regard to the labour market, the need to apply for all kinds of authorizations to build and operate research and production facilities, and a less dynamic market. The German government fears that enterprises will move their research facilities to other countries. This would harm the national networks that have developed between research facilities, producers and suppliers of capital goods. Germany feels its position threatened by countries close by that offer excellent research infrastructures, such as Sweden, Denmark and Switzerland, but also by countries further afield, such as Taiwan and South Korea (BMBF, 1998b).

Increasing Efficiency

The fact that the German government decided to extend the West German system of innovation to the *neue Länder* is an indication that it is basically satisfied with the structure of the national innovation system and confident about its merits. However, within the confines of this structure it is felt that some improvements are needed to let the knowledge system contribute more to the competitiveness of the German economy. More effective knowledge transfer from the public to the private sector is an important focus for policy. At present, though, the economic recession and the costs of reviving the economy in the east do not leave much budgetary room for new initiatives. Given static budgets, solutions are sought in efficiency gains. Rigid funding schemes for research institutes will be partly replaced by funding on the basis of specific programmes and the performance of research institutes. Consequently performance measurement and continuous evaluation will become increasingly important.

Bureaucratic and administrative rigidities will be removed in order to increase the independence and autonomous decision making power of research organizations. More government funds are to be allocated using tender mechanisms. The Helmholtz centres have transferred part of their funds to a central fund for strategic research that is under the control of the central board to finance specific programmes. Part of the federal institutional funding of the Blue List institutes has been transferred to DFG to be allocated through competition between projects put forward by different institutes. The *Fraunhofer-Gesellschaft* has been given permission to set up technology transfer offices and branch offices in the US and South East Asia. Helmholtz and Blue List institutes can retain their profits. Federal and state governments have agreed on further reforms that put the governments at a greater distance from the institutes. Extensive evaluations by international experts will be carried out to measure the performance of public research institutes.

Creating Favourable Conditions

The availability of a public research system that is marked by high quality and efficiency is already one important condition that should be favourable to innovation. However, there are more elements to an environment conducive to innovation. Among the issues that are being discussed are: the development of an information system; stimulation of the commercial exploitation of advances in knowledge; deregulation of the economy and tax reform; and the provision of venture capital. The creation of an up-to-date information and communication infrastructure demands significant investments in hardware. Technology transfer and exploitation is a worry in Germany as it is anywhere else. Policies to stimulate private R&D are thought to contribute to improving exploitation. Deregulation is a complicated issue, demanding an analysis of the way present regulation might unnecessarily hamper scientific and technological development. Some changes have been proposed to simplify the procedures for granting permission for investments in research and production facilities. It has already been decided that to establish some (mostly smaller) research facilities with no environmental impact official permission will no longer be needed. In tandem, the tax system is being reformed in order to improve conditions for the establishment of new companies.

A big issue is the lack of capital (in whatever form: venture capital, equity or loans) for SMEs and high-tech start-up companies. The data seem to confirm this: in 1995 only 20 new German high-tech companies were launched on the stock exchange, while 846 were floated on the American and 184 on the British stock markets. Rather than participating in companies itself, the federal government stimulates private investors to participate in SMEs and high-tech start-up companies. One way to do this is by deregulating the

German capital markets. It is proposed to change the regulations to allow institutional investors such as pension funds and insurance companies to invest in equity, and to allow holding companies more freedom to reinvest their profits. Furthermore, the government actively promotes participation in new enterprises through the provision of low-interest loans to venture capitalists and by taking over part of their liabilities. In this way, the government tries to involve private banks, participation companies and other investors more in venture capital markets. Finally, initiatives are being developed to improve the links between research institutes and venture capitalists. The objective is to offer venture capitalists more information on the basis of which to judge applications for contributions to SMEs and start-up companies. The federal government claims that its present proactive policy towards venture capital is quite successful. In 1996, 64 per cent of all European seed-capital funding settled in Germany. In biotechnology the number of companies has doubled within one year.

Reforms of Higher Education

The present German system of higher education has a number of weaknesses. In German universities, there is no selection for entry; a high school degree (*Abitur*) is sufficient to gain access to the system. While students are engaged in their studies there is very little personal guidance or monitoring of progress. There are only intermediate exams after two years and finals after five years (in theory at least). University education itself is free and a system of scholarships is available to those who need it. The consequence of all this is that German students in general take a long time to obtain their degree. The average student who gets his or her *Abitur* at 19 receives a university degree only at 28. Those who continue for a PhD are in their early thirties by the time they complete it. This leads to late entry of students into the workforce. Another problem with the German system is felt to be its one-sided academic character and its remoteness from the world of day-to-day business. German students are too little exposed during their studies to practical work experience in enterprises. Many students spend some time in a job before starting their studies, but this retards their entry to the labour market with a university degree even more. Finally, it is seen as ever more of a problem that the German educational system and its degrees are not compatible with the dominant systems and degrees elsewhere in the world.

These problems have led to a programme of reform of the educational system. The Kohl–government planned to reform universities with regard to the funding mechanisms, the selection of students, the appointment of staff, and the content and length of study programmes. Criteria of performance will be the basis for funding in the future rather than the number of students registered. This implies that a system to monitor and evaluate performance must be put in place. Monitoring and evaluation will involve research and teaching activities and both staff and students. Results of these evaluations will be made public. The federal government charged the Science Council with collecting the necessary data to carry out this monitoring of universities. If performance is to be the basis of funding, it is felt that universities should be given more instruments to enable them to improve their performance; therefore it has been decided to increase the autonomy of universities. For example, universities will be given the right to put into effect a selection procedure of their own to fill a certain share of the places on courses for which the number of applicants exceeds the number of places available. There are still quite a few reservations, though, regarding legal limitations on the number of years students are allowed to spend on their studies. Reforms are also foreseen in the systems of support and guidance offered to students, which will be more intensive, and in the frequency of examinations. Degree programmes should be reformed in such a way as to become internationally comparable. In the selection of staff more weight will be put on teaching capabilities. These measures should help to reduce the average duration of studies. At the time this book was written, it was not yet known whether the new government would continue to carry out the reforms that were initiated by the previous government. The Social Democrats opposed reforms of higher education for a long time.

Discussions are going on about changing the present system of human resource management to increase the attractiveness of a university career for young researchers. At present only professors enjoy tenure and have a great deal of autonomy in determining the activities of their research group. Even if they underperform, it is very difficult to discharge professors. As far as research is concerned, it is felt here too that exploitation of results should be higher on the agenda and that accessibility of knowledge should be improved. Probably this will encourage greater importance being given to applied research in universities.

The reform of the higher education and research system has led to intense discussions between the federal government and some state governments without whose consent reforms are practically impossible. State governments seem to object to these reforms for various reasons. They consider involvement of the federal government in this matter an attempt to intrude on their responsibilities, or they fear that the reforms might lead to the closure of universities in their area. However, Bavaria has already announced significant changes in its higher education system. These changes concern the introduction of competition among universities for government funds, and the reinforcement of the administration of universities. The president of the university, who is appointed by the Minister of Education, will see his powers extended and a university council (*Hochschulrat*) will be installed at each university to monitor the university on behalf of the Bavarian government.

Research Programmes

The contents of the R&D programmes that receive government support are under constant review. The federal government launched a number of ambitious initiatives in the areas of information technology, biotechnology and technologies deemed important for the service industries. Ample publicity, elaborate reports, the installation of high-level committees and the establishment of new organizations to promote these subjects seem to illustrate the weight that these themes carry in present federal R&D policy. It is not in the choice of themes that Germany seems to differ from other countries, but in the approach that is chosen. Besides increasing efficiency and the impact of publicly funded R&D, the development of adequate legislation and the creation of a general awareness of the issues at stake among the public are explicit policy objectives as well. This point is illustrated, for instance, by the way the Council for Research, Technology and Innovation operates, which with its reports paved the way for important government decisions on information and communication technology and on biotechnology. In its reports, the Council focuses on three aspects: (a) research, technology and application; (b) the legal framework; (c) social and cultural challenges.

An emphasis on legal, regulatory and social aspects in matters like ICT and biotechnology seems especially warranted in the present German context, where the public is generally critical about new technological developments. This is illustrated by the fact that the Environmentalist party plays a significant role in German politics and has become, more than in other European countries, part of the political establishment.

THE COUNCIL FOR RESEARCH, TECHNOLOGY AND INNOVATION ON ICT AND BIOTECHNOLOGY

The first report of the Council for Research, Technology and Innovation, released in 1995 after extensive discussions, was on information technology. It starts from the viewpoint that information technology has only just begun fundamentally to change our society. With regard to policy, the Council concludes that the emphasis should be on 'developing meaningful fields of application of the information and communication technologies'. The government should help to disperse information and to bring innovators together. R&D should be stimulated both on basic technologies and on technology application. The Council suggests emphasizing branch-specific solutions and the use of model applications. Concepts should be developed to introduce new technologies at SMEs.

The Council suggests a number of changes to the legal framework. In the case of new inventions, the government should impose regulation only after a pilot phase (and not before, since that would hamper innovation). The communication market should be deregulated almost completely. What is also needed is an adaptation of existing regulations, for example on anti-trust or intellectual property, taking account of these new technologies. For example, copyright and related property titles must be preserved when digitalization takes place. Other suggested changes concern the encryption procedures for documents in open networks (to be secured by independent, licensed 'trust centres') and the juridical use of electronic documents. Finally, the Council calls upon federal and state governments to invest heavily in the introduction of new applications of ICT in schools and universities.

The second report of the Council for Research, Technology and Innovation (1997) concerned biotechnology. Fearing that Germany might miss important opportunities in this field, the federal government increased its R&D budget on biotechnology by almost one-third over a period of five years. The presentation of the final report of the Council was accompanied (or even preceded) by a large number of initiatives. The Council pleads for better co-operation between industry and science by expanding existing networks and platforms. Also it argues for improving the conditions for starting up new companies and for industrial research, and for the concentration of government funding in particular areas. Of interest are the recommendations that concern the ethical and legal aspects of biotechnology applications. The Council holds that the rules for starting research projects in biotechnology under the Genetic Engineering Act and other acts are far too complicated. Interpretation of the rules by various authorities seems to be arbitrary: simplifying them can be accomplished without sacrificing vital safety standards. Finally, the Council advises the establishment of 'a suitable interdisciplinary body closely associated to the government' to monitor and evaluate R&D and to bring out reports on basic developments taking place in biotechnology.

German R&D Policy in its International Context

Due to the size of the German national system of innovation, the need to co-operate internationally is less pressing than in smaller countries like the Netherlands and Finland. However, the general attitude towards co-operative R&D in the EU has always been one of firm support. Germany defines its R&D policy objectives clearly within a European and global context. In fact, we could describe these objectives as follows (BMBF, 1993):

- to secure, in the globalization process of industry and science, Germany's position, especially in the high-technology sector, as a contributor to research and production in Europe and as part of the triad formed by South East Asia, the US and Europe, thus making an essential contribution to growth and employment in Germany itself;
- to further resolutely the economic and political union of Europe. The European research and technology community is an essential element of a policy designed to secure Europe's political, cultural and economic strength in international competition in order to ensure peace and prosperity;
- to offer the neighbouring countries in the East the assistance and partnership of the European community in order to permit the development of free and democratic structures, also in science and industry;
- to make an effective German contribution to solving global problems in the areas of energy supply, climate development and environmental protection (with as main themes: global warming, the ozone hole, tropical ecology); also with regard to the developing countries, to aim at sustainable growth without progressive resource consumption.

Joint policies on R&D within the EU are not only seen as instruments to promote science and technology and to facilitate research spillovers between countries, but also as one of the instruments to foster cohesion within Europe. Germany is strongly aware of its responsibilities in shaping the agenda on international collaboration in R&D. This applies to the contents of EU Framework programmes as well as to technological help given to the countries of Central and Eastern Europe. In Germany, the discussion about EU R&D policies tends to revolve around their content rather than the size of the budget. However, the recent signal of the German government that it feels that its contribution to the EU is too large will have repercussions on its policy stance regarding R&D budgets. More emphasis will be put on efficient use of available resources. However, additionality and the national return on investments in European R&D are not prominent themes in the discussion. Here, the view is taken that R&D policy as such is just a part of the whole of community policies. The payoff to co-operation in R&D policy may well come in some other area, for example, social or agricultural policy. However, R&D is also a field where the principle of subsidiarity is considered important. Too much centralization would result in a loss of variety, to the detriment of the quality of research.

The federalist structure of Germany is reflected in German representation at the EU. Like other countries, Germany has two representatives in CREST (*Comité de la Recherche Scientifique et Technologique*), one coming from BMBF, the other from the Ministry of Economic Affairs. In addition, however, the states each has its own representative with observer status. The determination of the German position in EU matters regarding science and technology policy takes place along the standard procedures within the government, but for this subject it acts in consultation with the states. Frequently there is also a dialogue on EU policy matters between BMBF and the different institutions carrying out R&D.

6.4   POLICIES

In 1996 the private sector received about 28 per cent of total federal expenditures on R&D. This amount equals between 8 and 9 per cent of the private sector's total R&D expenditures. More than 60 per cent of these expenditures relates to R&D that is carried out by private companies for the government, mainly military research. Less than 40 per cent is used to stimulate the private sector's own R&D activities. This happens either indirectly through subsidizing private non-profit organizations that carry out or fund research for SMEs (such as AiF) or directly through subsidies for private companies, mainly SMEs. Particular policies concern the employment of R&D personnel, the establishment of technology-based companies and the co-operation between SMEs and research institutes. Most of these direct subsidies are only available to SMEs in the new states; SMEs in the old states also receive financial support, but to a far lesser extent. Rather than providing direct subsidies to SMEs the federal government tries to develop an infrastructure that helps them to innovate. An example is the development by BMBF of an electronic service to ease access of SMEs to government institutions and support programmes. EFA (*Service Elektronischer Förderantrag*) enables SMEs to apply for R&D support in an easy and cheap way. Its introduction has significantly shortened application and decision making procedures: four weeks now against more than eight in the past.

Table 6.4 gives an impression of current federal expenditures on R&D. Basic funding of DFG, the *Max-Planck-Gesellschaft* and the *Fraunhofer-Gesellschaft*, plus a number of specific expenses on universities (most of the costs of univer-

Table 6.4   Federal R&D expenditures in 1998

	Expenditure (DM million)	Share (%)	Growth 1996–8 (%)
Basic funding of DFG, Max Planck, Fraunhofer and universities	3 130	19	5
Defence research	2 744	17	–5
Space research	1 427	9	–8
Environmental research	1 068	7	1
Large equipment for fundamental research	1 036	6	–3
Information technology	986	6	–5
Energy research	845	5	3
Health research	791	5	4
Improvement of conditions for innovation and technology transfer	759	5	–9
Other	3 446	21	–6
Total	16 232	100	–2.6

Source: BMBF (1998a).

sities are borne by the states), consumes together about one-fifth of the federal R&D budget and its share is growing. With a share of 17 per cent, defence research is the second-largest category in the budget; all other categories are groupings of research programmes in specific fields and are considerably smaller. The federal budget for the stimulation of innovation and the improvement of the innovation environment is decreasing sharply. This is mainly due to the expiration of a development programme of the Ministry of Economic Affairs. Of the smaller programmes, the budgets for biotechnological research and for terrestrial transport systems have been growing at double-figure rates over the last two years. About 10 per cent of the federal expenditures on R&D go to international research organizations and programmes (including organizations like CERN); this share is decreasing, but only slowly. In 1998, the planned overall federal government budget for R&D amounted to DM16.2 billion whereas expenditures were DM16.7 billion in 1996. The total decrease of about 2.6 per cent is not compensated by an increase in spending by the states.

Leitprojekte

To increase efficiency of research and to shorten the period between discovery and commercialization of research results the federal government set up the so-called *Leitprojekte* (lead projects). *Leitprojekte* concentrate means at specific areas of interest and stimulate collaboration between research institutes. The procedure is straightforward. The government, advised by universities, enterprises and research institutes, selects a number of themes. Once a theme is chosen, competition for funds is open to groups consisting of enterprises, research institutes, universities and governmental organizations. In the research proposals it is indicated how collaboration between partners will take place and how the results will be commercialized. Partners are expected to bring in significant resources of their own. In the first round of selection an independent jury judges the proposals and selects at most 15 of them. The consortia that submitted the proposals are asked to elaborate their proposal. The costs of elaborating the proposal will be borne by BMBF. In this phase other partners, especially SMEs, can join the consortium. Finally, the jury selects at most five projects that will receive significant funding. *Leitprojekte* cover broad themes such as 'Innovative products based on new technologies'; 'Use of internationally available knowledge for education and innovation'; 'Mobility in agglomerations'. In March 1998 the first proposals were approved, so the long-term effects of the *Leitprojekte* are not yet known. Yet the government thinks that the procedure will lead to the creation of new, promising partnerships, even among those whose proposals were not accepted.

Patent Initiative

The government feels that SMEs and universities do not protect their intellectual property sufficiently. To increase the number of patents applied for, the government has launched several initiatives. SMEs will be subsidized when they apply for a patent. SMEs either do not have to pay the full costs of the patent application or they receive a cheque that can be exchanged at the Patent Advisory Centres. Research institutes and universities are stimulated to develop patenting strategies. The idea is that every research result that is commercially applicable should be patented, and that income from patents should flow to the individual research groups and researchers. SMEs that have not applied for a patent over the last five years, but are willing to do so in the future, will receive financial and practical aid from organizations that are concerned with patenting. The aim is to help companies to apply for a patent for the first time and to use patent information. Students at universities and *Fachhochschulen* will be taught how to deal with patents and how to use information on patents. Finally, in order to improve the access of SMEs to information sources in general (not only patents) the government created an interactive databank that can be accessed by SMEs.

Regional Contests

The *Biotechnologie 2000* programme supports all kinds of research on bio-technology. An interesting point about this programme is the way in which it was launched. BMBF organized the so-called *Bioregio-Wettbewerb*, a contest aimed at stimulating regions to come up with plans to create an integrated system of scientific and commercial activities around a specific theme. Then BMBF selected a few regions which offered the best plans. For the near future, financial means for biotechnology will be concentrated in those regions. By concentrating its means geographically, BMBF hopes to stimulate the efficient use of funds and local research capacity. BMBF appears quite content with this way to launch a programme. It concludes that even in those regions that were not selected in the end, different interest groups were brought together and came to take initiatives that might not have been taken without this contest. The results and spin-offs of the *Bioregio* contest are considerable. By the spring of 1998, about 150 new companies in biotechnology had been established. The programme was also successful in mobilizing private capital. At the beginning of 1998 about DM570 million was available. Requests for financial support from BMBF amounted to DM115 million.

Due to its success in biotechnology, regional contests were also launched around other themes. The main criteria by which regions and their proposals are judged are:

- the number, profile and capacities of the research institutes, universities and *Fachhochschulen* located in the region;
- the extent and quality of the existing interdisciplinary network in research on the theme subject of the contest;
- the number and characteristics of established industrial companies and service providers (patent offices, data networks, technology and other advisers);
- the use of existing scientific theme-related know-how and regional research capacities for the development and marketing of new products, production processes and services;
- the policy measures that have already been taken in support of the theme figuring in the contest, and the establishment of new theme-related companies;
- the willingness of banks and other capital providers to invest in regional theme-related companies;
- the extent of collaboration between organizations;
- regulations pertaining to the establishment of theme-related production facilities.

Dienstleistungsinitiative (service sector)

The dynamism in the service sector is the main driving force behind job creation. The technology-orientated service sector shows much more dynamism than large parts of manufacturing and the number of newly established technology-based service enterprises is much larger than the number of new industrial enterprises. The high-tech service sector is not only a motor of job creation, but also a sector with a crucial impact on the competitiveness of manufacturing industries. The local presence of a well-functioning, up-to-date service sector with high-level capacities in functions such as design, finance, planning and marketing is considered an important determinant of the location decision of industrial firms. BMBF notes that the service sector is highly fragmented (few linkages among different actors, less organized) and is less experienced with regard to R&D as compared to the manufacturing sector. Creating linkages therefore was the first thing to do. For that reason the federal government launched the *Dienstleistungsinitiative*, which offers a platform for research and discussion on the development of the service sector and is supposed to stimulate new initiatives by individual participants. Now that the initiative has proved to be a success (at least the number of participants is growing) the objective is to establish a sound and increasing network of research institutes, government organizations and enterprises with clear common goals such as to influence political decisions. A concrete policy measure that evolved from the *Dienstleistungsinitiative* is the establishment of an electronic market place for discussion and information exchange.

SERVICES 2000+

The research project *Services 2000+*, which brought together firms, universities and research institutes, looked at the perspectives and requirements of the service industries in the next millennium. In May 1998 the conclusions and recommendations that resulted from this project were made public. The main recommendations were:

- to adapt the educational system and the choice of degree courses to the needs of the new service sectors, especially covering the use of new media and the knowledge of other cultures (for example, the Turkish culture, as there is a large Turkish minority in Germany);
- to reorientate research programmes and scientific developments which are currently orientated mainly towards

meeting the needs of manufacturing more in the direction of the service sector;
- to improve patent protection on service innovations in order to speed up innovation processes in the service sector;
- to improve statistical data on services;
- to encourage the establishment of new labour relations, the optimization of business relations (*Geschäftsprozesse*) and changes in human resource management.

After the results of this research project were published, the federal government took a number of policy initiatives, such as to develop markets for new types of services and to improve the organization of service enterprises. The total volume of these measures is DM38 million.

Länder R&D Policies

Science and technology policy is a shared responsibility of the federal government and the 16 German *Länder*. The *Länder* each have their own R&D policies that are partly independent of federal policy. The states are quite diverse in size, economic structure and level of development, and they spend different amounts on R&D. Table 6.5 shows the percentage of total German R&D performed in a specific state (financed by the federal government, the state government and the private sector), and the share of this total financed by the state government. The table also shows the share of total German GDP per state and the ratio of R&D expenditures and GDP per state (including federal R&D expenditures that flow to a specific state).

Three states, Baden-Württemberg, Bavaria and Northrhine-Westfalia, take account of 60 per cent of all R&D activities by the German states. Of these three, Bavaria and Baden-Württemberg have a share in R&D that is significantly above their share in GDP. These states host a relatively large number of research institutes that are financed jointly by the federal and the host-state governments. Bavaria and Baden-Württemberg are among the wealthiest German states, with low unemployment rates and high average incomes. Only Hessen and Hamburg have higher average incomes, but unemployment rates are somewhat higher there. Northrhine-Westfalia, the largest German state, is still recovering from the decline of its smoke-stack industries in the *Ruhrgebiet* that took place in the 1970s and 1980s. The new states (marked with [a] in Table 6.5) have unemployment rates are between 15 per cent and 18 per cent, while average incomes are only 50 per cent of the German average. However, eight years after unification R&D activities have reached levels that start to

Table 6.5   Shares of GDP and R&D expenditures by state in 1994

State	Share of national R&D performed in a state (%)	Share of GDP generated in a state (%)	R&D expenditures as a share of state GDP (%)	Funding by state government as share of R&D performed in state (%)
Baden-Württemberg	23	14.4	3.8	11
Bavaria	20	16.9	2.8	13
Northrhine-Westfalia	17	22.4	1.8	20
Hessen	9	9.8	2.1	14
Lower Saxony	7	8.9	1.7	22
Berlin	6	4.3	3.3	28
Rheinland-Pfalz	4	4.4	1.9	18
Hamburg	3	3.9	1.7	21
Schleswig-Holstein	2	3.2	1.2	34
Bremen	1	1.2	2.8	18
Saarland	1	1.2	0.9	51
Saxony[a]	3	3.0	2.3	37
Brandenburg[a]	1	1.8	1.5	36
Mecklenburg-Vorpommern[a]	1	1.1	1.2	23
Saxony-Anhalt[a]	1	1.8	1.5	45
Thuringia[a]	1	1.6	1.7	45
Germany	100	100	2.4	18

Note: [a] New states.

Source: BMBF (1996).

match their share of GDP. The share of total R&D expenditures paid for by the state governments differs between states. In general, the governments of the new states pay for a larger share of R&D than the other state governments. If federal funding of local R&D activities is added to this, it turns out that in 1994 at least 70 per cent of all R&D activities in the new states was financed by government. This was significantly above the German average of about 37 per cent.

Example 1: Hessen

Hessen is one of the wealthiest and most prosperous states of Germany, producing about 10 per cent of German GDP with 7.5 per cent of the German population on 6 per cent of its territory. It accounts for 9 per cent of German R&D activities, of which three-quarters is financed by the private sector. In 1994, manufacturing industries accounted for about 37 per cent of employment, half of which was generated by chemicals, motor vehicles, electronics and machinery. Services are also very important, Frankfurt am Main being one of Europe's financial centres and one of its main airports. To remain competitive in manufacturing and services, Hessen has an extensive public knowledge infrastructure: five universities, five *Fachhochschulen*, four Blue List institutes, one main establishment and three branch offices of the *Helmholtz-Gemeinschaft*, nine institutes or research groups of the *Max-Planck-Gesellschaft* and two establishments of the *Fraunhofer-Gesellschaft*.

In Hessen, the Ministry of Science and Cultural Affairs and the Ministry of Economic Affairs, Transport and Spatial Planning are jointly responsible for R&D policy. Their policy is supported by a number of organizations. The HLT group is the official economic development organization of Hessen. It consists of two companies operating under joint management. The first, the economic development bank of Hessen (*Wirtschaftsförderung Hessen Investitionsbank AG*), carries out the government's financial R&D support policies for private industry and offers information and advisory services to domestic and foreign companies. The second, the Company for Research, Planning and Development (*Gesellschaft für Forschung, Planung, Entwicklung mbH*), acts as a policy research, planning and development institution which advises state, regional and local administration. The *Forum Wissenschaft–Wirtschaft* is a platform that brings government, enterprises and scientists together. It consists of representatives of the government, the chambers of commerce, employers' organizations, universities and *Fachhochschulen*. The *Technologie Stiftung* stimulates innovation and technological collaboration among SMEs and the various technology transfer organizations. Furthermore, it brings together companies, technology suppliers, intermediaries and policy makers, and tries to create awareness of technological developments among SMEs.

The general objective of Hessian science and technology policy is to maintain the competitiveness of the Hessian economy through innovation and thereby to secure employment in the long term. Environmental protection is another priority of the state's policy. The government explicitly states that it wants to use economic instruments to realize these objectives, especially with regard to the implementation of environmental technologies. The government helps to reduce the risks to the entrepreneur but does not take over responsibilities or initiatives. Nevertheless, government offers some financial support to individual companies and stimulates the supply of venture capital. If direct financial support is offered, this is usually directed at the launch of new companies, the gathering of information by SMEs or the decrease in environmental pollution. However, the emphasis is on non-financial instruments. A general aim is to help companies to convert scientific knowledge and technological know-how faster into new products and services. Therefore attempts are made to establish and to extend a technology transfer infrastructure that is closer to the private sector, with well-developed technology service providers, applied research institutes and co-operative networks. To establish these organizations, financial involvement and managerial commitment is sought from firms and universities. New organizations usually get a start-up subsidy but have to be financially independent from the government after a few years. So far, the Hessian government has encouraged the establishment of many intermediaries within a wide variety of organizations, such as *Industrie- und Handelskammer* (Chamber of Commerce), *Handwerkskammer* (Chamber for Handicraft), labour unions, employers' organizations and branch organizations.

Example 2: Saxen-Anhalt

Saxony-Anhalt is one of the *neue Länder* and one of the poorest states. With only 2 per cent of German GDP produced by 3.5 per cent of the German population on 6 per cent of its territory, it accounts for no more than 1 per cent of German R&D activities. These activities are directed mainly towards improving the competitiveness of existing products and production processes. In 1994, less than one-quarter of all R&D activities in Saxony-Anhalt were financed by the private sector. Saxony-Anhalt inherited from the former GDR an outdated production structure and a severely damaged natural environment.

Its social and economic situation is considered to be the worst in Germany: it has the lowest per capita income and the highest unemployment. After unification total employment decreased by 37 per cent. Employment in manufacturing halved and agriculture was hit the worst with a decrease in employment of 80 per cent. Employment in 'other services' more than doubled. Even today only a small number of newly established firms are in the industrial sector. Currently, the economic situation has stabilized, but problems remain. Wage increases and transfer payments from the federal government have boosted the development of companies which are mainly producing for local markets. In these markets they are protected against competition by high transaction and transport costs. Export-orientated enterprises that are subject to international and national competition have scarcely developed.

For Saxony-Anhalt, the main objective of science and technology policy is to catch up with the rest of Germany, both by stimulating private sector R&D

and by developing the public knowledge infrastructure. Currently, there are three universities and five *Fachhochschulen*, some of which do not cover technical disciplines, however. Over the last couple of years, the *Max-Planck-Gesellschaft* and the *Helmholtz-Gemeinschaft* have each established institutes in Saxony-Anhalt, five Blue List institutes have been founded and the *Fraunhofer-Gesellschaft* has opened two establishments. Some of these institutes are clearly future orientated, but others focus on the present problems of Saxony-Anhalt, such as the Fraunhofer Institute of Manufacturing and Automation and the Helmholtz Centre for Environmental Research. Apart from the public institutes there are several research centres that used to be part of large GDR enterprises, but that have since been outsourced and privatized. This is a highly diverse group of institutes that until recently received no structural support from the government (neither federal nor state). The state government wishes to retain them, but the fact that so far they have been unable to raise enough private funding casts some doubt on the perspectives of these institutes.

In Saxony-Anhalt the Ministry of Economic Affairs is responsible for technology policy. It supports business-orientated research, R&D projects in private companies and the build-up of a technological infrastructure. The Ministry of Education, Science and Culture (*Kulturministerium*) is responsible for education and research at universities and research institutes. There exists a large and highly diversified group of intermediary organizations. Most of them are concerned with technology transfer, partner search for SMEs, advice on financial and technological affairs, project management, the organization of training courses and workshops, and the like. Some concentrate on specific themes (for example, environmental technologies, energy), others offer more general services to companies.

As in other states, commercial use of new technologies is here also a major focus of policy. Increasing budgets are made available for co-operative research projects, pilot and demonstration projects and the support of technology diffusion. Public services and infrastructure for SMEs and innovative start-up companies are to be improved. This implies a complete restructuring of the present intermediary structure which is too diversified and confusing. The number of transfer agencies will be drastically reduced, and their non-government funding must be at least 50 per cent. With regard to R&D content, government support will focus on specific technological areas such as biotechnology, material sciences, neurosciences, production systems and environmental technologies. The objective is to create a system that supports all stages of innovative activities from invention, through the design of a product or process to market introduction. A stable policy environment is to be created such that companies have the possibility to take account of government support policies in their plans. Most support policies of the government of Saxony-Anhalt are supplementary to that of the federal government, the only exception being the provision of venture capital. The state of Saxony-Anhalt founded the *Innovations- und Beteiligungsgesellschaft* (IBG), a company that participates in innovative projects.

6.5 CONCLUSIONS

Over the past few years the German economy has had to meet a number of major challenges. The economy of the *neue Länder* started to recover only slowly after its almost complete overhaul at the beginning of the 1990s. The rest of the German economy was hit by recession. Under these circumstances, the overriding feeling seems to be that the science and technology system functions fairly well. Certain weaknesses identified in the German industrial structure, a certain conservatism reflected in the opinion that German industry tends to stick to its knitting, are addressed through a more direct involvement of the state in these areas (the state has, for example, initiated investments in Aids research). The innovation system proved able to deal with the challenges of unification; the new states have received special attention. This has taken mainly two forms: the rapid creation of new outlets for the Federal Republic's research organizations; and special support policies for SMEs. However, the radical reorganization of the GDR science base has had a considerable cost in human capital.

Universities, *Fachhochschulen* and a large number of state-sponsored research institutes constitute the backbone of German public R&D. On the education side, an important change taking place is the reorganization of university education, bringing it more into line with systems in other countries. The duration of degree courses is being shortened and the guidance of students intensified. On the research side, a main tendency is to improve the efficiency of research processes and the exploitation of results. To increase efficiency, the government has started to concentrate financial means in specific geographical areas, to stimulate co-operative research, both nationally and internationally, and to build incentives into the research system to make it more market orientated. In the future, research institutes and universities will have to compete for funding more often. The German government is comparatively reluctant to give direct financial support to R&D in companies. The main thrust of policy is on the maintenance of a high-quality public knowledge infrastructure and on creating conditions that stimulate innovativeness.

The federalist structure of Germany has both advantages and disadvantages for the management of the knowledge infrastructure. On the one hand, it gives science and technology policy a certain flexibility, enabling the states to react to the particular circumstances with which they are confronted. On the other hand, it involves complex decision making procedures and can lead to significant retardation in adapting to changing circumstances. This is most prominently illustrated by the continuous debate on the reform of the education system.

REFERENCES

Bauer, G. and J. Zabel, 1994, *Technologiebericht Hessen 1994*, HLT Report Nr. 426, Wiesbaden: HLT Gesellschaft für Forschung, Planung, Entwicklung mbH.

Bundesministerium für Bilding, Wissenschaft, Forschung und Technologie (BMBF), 1993, *Bundesbericht Forschung 1993*, Bonn: Bundesministerium für Bilding, Wissenschaft, Forschung und Technologie.

——, 1996, *Bundesbericht Forschung 1996*, Bonn: Bundesministerium für Bilding, Wissenschaft, Forschung und Technologie.

——, 1998a, *Bundesbericht Forschung 1998*, Bonn: Bundesministerium für Bilding, Wissenschaft, Forschung und Technologie.

——, 1998b, *Zur technologischen Leistungsfähigkeit Deutschlands: Endbericht an das Bundesministerium für Bildung und Forschung*, Bonn: Bundesministerium für Bilding, Wissenschaft, Forschung und Technologie.

——, 1998c, *Reformen 1994 bis 1998: Wissen + Innovation = Arbeit*, Bonn: Bundesministerium für Bilding, Wissenschaft, Forschung und Technologie.

Council for Research, Technology and Innovation, 1995, *The Information Society: Opportunities, Innovations and Challenges*, Bonn.

Council for Research, Technology and Innovation, 1997, *Biotechnology, Genetic Engineering and Economic Innovation: Making Responsible Use of Existing Opportunities*, Bonn.

Kemp, R., I. Demandt and B. Dankbaar, 1996, *Monitoring Technology Policy in Europe: The Role of Public Research Institutes, A Study for the Dutch Ministry of Economic Affairs*, Maastricht: MERIT.

Ministerium für Wirtschaft und Technologie des Landes Sachsen-Anhalt, 1995a, *Jahresbericht 1995: Fakten & Projekte*, Magdeburg: Ministerium für Wirtschaft und Technologie des Landes Sachsen-Anhalt.

——, 1995b, *Technologieführer Sachsen-Anhalt*, Magdeburg: Ministerium für Wirtschaft und Technologie des Landes Sachsen-Anhalts.

——, 1996a, *Stand der Industrieforschung, der industrienahen Forschung und der Technologieentwicklung in Sachsen-Anhalt: Beschluß des Landtages vom 31. Mai 1996*, Magdeburg: Ministerium für Wirtschaft und Technologie des Landes Sachsen-Anhalts.

Landestag Sachsen-Anhalt, 1996, *Stand der Industrieforschung, der Industrienahen Forschung, und der Technologieentwicklung in Sachsen-Anhalt, Beschluß des Landtages*, 31 May, Magdeburg.

Riel, C.J. van, 1995, *Technologiebeleid in enkele Europese regio's; zeven reisverslagen*, AWT, The Hague, Achtergrondstudie nr. 7.

7.   Five non-European countries

7.1   INTRODUCTION

In this chapter we briefly discuss science and technology policy in five non-European countries: Israel, Japan, Singapore, Taiwan and the US. These countries have been selected either because of their economic importance or because of their excellent technological and economic performance over recent years. Whereas we gathered material for the chapters on the main case study countries from interviews with civil servants as well as written sources, we use only published sources of information for the countries covered in this chapter. For the section on Japan we draw also on some discussions with academics. Each section comprises of a short introduction of the country, a description of the existing science and technology structures, a note on the philosophy behind a country's science and technology policy and an overview of policy measures.

7.2   ISRAEL

Introduction

No analysis of the Israeli economy can be made without reference to the Palestinian conflict and the continuous hostilities with its neighbours. As noted by JIIS (1993), for example, the need to ensure a high deterrent capacity strained the economy significantly and limited its long-term growth potential. This became evident during the 1970s and 1980s when Israel was confronted with hyperinflation (445 per cent in 1984) and economic stagnation. In the 1990s the economy was boosted by a major influx of immigrants from the former Soviet Union (an estimated total of 1 million by the end of 1995) and the peace process initiated by the Labour government headed by Prime Minister Rabin. GDP grew at an average rate of 5.9 per cent over the period 1992–5. GDP per capita grew at a more modest rate of 3.1 per cent, while unemployment decreased from 10.6 per cent in 1991 to 6.3 per cent in 1995. R&D expenditures equalled 2.3 per cent of GDP in 1992, of which about 56 per cent was spent on industrial development (R&D on agriculture, fishing and forestry accounts for 14 per cent of government R&D expenditures). The Israeli economy is strong in high-tech products. Electronics, chemicals, and perhaps biotechnology are at a high level. Israeli high-tech firms have been successfully introduced on the National Association of Security Dealers Automated Quotations (NASDAQ). Furthermore, Israel has the highest percentage of engineers and university graduates in the world (more than twice that of the US). It should be kept in mind that a large part of this technological capacity is used by the military sector. This may have created a comparative advantage in commercialized military products and caused some spillover to the civilian sector. However, it has also crowded out the competitive capacity for other (civilian) products (Teubal, 1993).

Although the Netanyahu government states that there is no strong link between the peace process and economic performance, there are signs to the contrary. In the second half of the 1990s the Israeli economy has been stagnating again. In 1996, the state owned approximately 90 companies (subsidiaries included), of which 47 are termed commercial enterprises. It had a stake of less than 50 per cent of the outstanding shares in another 28 companies, ten of which are commercial enterprises. These companies accounted at that time for 17.5 per cent of national exports and 15.5 per cent of GDP. Privatization is deemed appropriate but the stock market is too unstable to support such a move on a massive scale. Moreover, the government deficit is substantial and monetary policy is very tight, reflecting fears of (hyper)inflation (*NRC-Handelsblad*, 30 November 1996). The flow of immigrants which boosted the Israeli economy for several years is now rapidly drying up. Whatever the claims about the relationship between the economy and the peace process, economic stagnation is threatening the R&D budgets of the government. After years of growth, budgets are now tending to stabilize or even to decline.

Structures

The Ministerial Committee for Science and Technology is responsible for the promotion, co-ordination and supervision of the science and technology activities of the civilian government ministries. It determines the long-term policy on science and technology. The ministerial committee is advised by the National Council for Research and Development, a public body consisting of scientists, industrialists and representatives of various economic and research sectors. The ministries most prominently involved in science and technology are the Ministry of Science and Arts, the Ministry of Trade and Industry and the Ministry of Defence. The last of these has a very special position.

Within each ministry the Office of the Chief Scientists is responsible for the formulation and implementation of policies regarding research and technology. The Ministry of Science and Arts chairs both the Ministerial Committee for Science and Technology and the Chief Scientists' Forum (a committee consisting of the chief scientists of all ministries). It is actively concerned with application-orientated research and manpower development, international co-operative agreements and the appointment of chief scientists at other ministries. In general the Ministry of Science and Arts seems to be concerned with basic and generic research. Applied research is the responsibility of the Ministry of Industry and Trade. The Office of the Chief Scientist (OCS) of this ministry formulates and implements policies regarding the development and expansion of technology-based industry. OCS chairs the Research Committee, which consists of civil servants and representatives of public research institutes. The committee approves R&D proposals and sets the terms under which grants and aid are allocated from the OCS budget to companies. OCS represents the government in the management of ten institutes dealing with industrial R&D. Of particular importance is the Ministry of Defence. Its decision structures seem to be largely separate from the civilian structures and it channels a significant portion of the government R&D effort to the private sector. The Ministry of Defence also has close links to the Rafael Armament Development Authority (6000 employees, of whom 2000 are employed in R&D) and the Israel Military Industries (14 000 employees) which carry out significant amounts of research.

There are ten universities (or institutes of comparable level) in Israel. They differ in scale, the variety of studies offered and scientific activities. The Hebrew University in Jerusalem, Technion in Haifa and the University of Tel-Aviv are by far the most important institutes of higher education. All Israeli universities host organizations to promote the commercialization of academic research. Adjacent to most universities, science-based parks have been founded. Some research and testing institutes for Israeli industry are hosted at universities. Basic research at the universities is funded by a large variety of organizations and foundations, such as the Council for Higher Education, the Israel Academy of Sciences and Humanities (which administers the National Science Foundation) and a large number of international collaboration programmes (*Nature*, 29 June 1995). The funding system is complicated and very fragmented as the government in turn funds many of these funding organizations.

The largest civilian research organization is the Industrial Research Organization (IRO) which consists of ten research institutes. The IRO institutes have been severely criticised: according to Kaufmann and Yinnon (1996), these research organizations (or technology centres, as they call them) hardly function, except the Plastics Centre (which is managed by the Rubbers and Plastics Association) and the Israel Institute of Metals. Kaufman and Yinnon are afraid that the lack of 'technology centres' will hamper the entrance of private companies into national and international research programmes. Unfortunately, nothing has changed yet under the Netanyahu government. Apparently the creation of new research organizations seems to be opposed by the Ministry of Trade and Industry: the Ministry is afraid of creating 'white elephants' that have high costs and do not meet market demands. Other industrial research institutes exist outside the IRO framework; most are rather small with two significant exceptions: the Agricultural Research Organization (ARO) consolidates nearly all the research activities of the Ministry of Agriculture; the Rafael, the Armament Development Authority is a civilian authority with strong links to the Ministry of Defence.

There exist several organizations in Israel that act as intermediaries between government, universities and research organizations, on the one hand, and the private sector, on the other. The most important of these organizations is the Israeli Center for Industry R&D (MATIMOP), which is said to be 'the implementing arm of the Office of the Chief Scientist' (MIT, 1996b). MATIMOP is a public, non-profit organization founded by three associations of manufacturers.

Philosophies

Israel has a strong tradition of government intervention in many industrial activities, especially R&D. Since the birth of the state of Israel, R&D has been substantial relative to GDP. In fact, investing in R&D and science might have been one of the few instruments to combine economic growth with a high deterrent capacity. During the 1950s and the 1960s government aid was mainly directed (apart from the defence sector) at the food, textile and metals industries. During this period, several research institutes and universities were founded. At the end of the 1960s, the government decided to engage in a more structured and co-ordinated approach. The science and technology efforts of different ministries were co-ordinated and the government established quite generous financial support for internal private R&D. Applicants had few obligations and there were no limitations on the subject of R&D. According to Teubal (1993), this type of support was appropriate for the then prevailing technological needs. Teubal (1993) suggests that the stagnation of the economy during the 1970s and the 1980s can be partly attributed to a number of serious defects in the Israeli national system of innovation, some of which still exist today. Financial support for R&D, instead of being selectively directed to high-tech sectors' co-operative research projects and to generic research, is too neutral and too universal. Most support is directed at research (generation of new knowledge), leaving the financing of the other phases of the innovation process completely to the private sector. Most of the existing research institutes (often supervised by the Industrial Research

Authority) are directed at traditional industries and do not succeed in fulfilling the long-term needs of small and medium enterprises SMEs. Finally, Teubal suggests that the R&D activities of universities and intermediary institutions are rather uncoordinated.

The philosophy of current science and technology policy partly meets the critics. To overcome the dilemma of being a small country, a sharp distinction has been made between science and technology. In science, Israel attempts to maintain a basic standard of academic expertise across a broad spectrum of scientific fields. In technology, it is concentrating on a few high-tech fields, especially computer software, microelectronics, optics and biotechnology. To correct the weaknesses mentioned above, technology policy will become more specific, and risk sharing and co-operation will be promoted. Another issue in science and technology policy is the integration of highly educated immigrants into Israeli R&D. For this purpose, a number of plans were developed. Support and incentive programmes are governed by the Law for the Encouragement of Industrial Research and Development 1984, which also reveals something about the philosophy of Israeli R&D policy. The general purpose behind the government's policy is to share in the risk inherent in industrial research and development. More specifically, the purpose of the aforementioned law is to support industrial R&D in order to 'foster the development of technology-oriented industry in Israel, while utilizing and expanding the existing technological infrastructure of Israel's industrial and scientific communities and of its human resources'. Israel's balance of trade must be improved by increasing the manufacture and export of high-tech products developed within Israel and reducing reliance on imported goods of this type (MIT, 1996a, b). The Israeli government strongly advocates international co-operation to reduce the costs and to share the risks of R&D activities. Furthermore, it hopes to gain access to foreign markets in order to commercialize Israeli products. To promote international co-operation the government has signed agreements with several other countries and the EU. These programmes are based on a number of principles, such as assistance in identifying partners, assistance with applications for aid, co-ordination between governments, and financing a portion of the actual R&D costs.

The present government is forced to cut back its expenditures due to the widening government deficit (caused by smaller than expected tax receipts). The budget for science and technology was expected to decrease by 5 per cent over 1997. However, to meet its own objectives, the government should have increased the science and technology budget by 17 per cent (*The Times Higher Education Supplement*, 26 July 1996). Other policy objectives of the present government are the privatization of state-owned enterprises, some of which are very knowledge intensive, and the reduction of tax barriers, thereby exposing industry to international competition.

Policies

The Israeli government offers a wide range of financial support schemes. Apparently, the schemes now also cover various stages of the innovation process, including such activities as prototype development. Specific support measures exist regarding the location of companies in science parks or backward areas. Of special interest are the Technology Incubators Programme and the MAGNET programme.

To use the potential of highly educated immigrants the Israeli government started to encourage them to engage in innovative commercial activities by launching the Technological Incubators Programme. This programme consists of centres or frameworks (incubators) that house groups of scientists (and projects). The scientists are supposed to develop new products or processes for commercial purposes. To enable entrepreneurs to build up a company which can stand on its own feet after some time, the programme offers funds, administrative assistance, professional guidance and physical premises. Meanwhile, there are 27 technological incubators throughout the country, acting as independent entities with their own managers and boards of directors.

As indicated above, main weaknesses in the Israeli R&D system in the beginning of the nineties were felt to be a lack of focus of R&D policy, a lack of support for knowledge exploitation and diffusion, and a lack of co-ordination between research institutes, universities and the government (Teubal, 1993). The counter these problems, the government started up the MAGNET programme in 1992. MAGNET stands for Generic Pre-competitive Technologies and R&D, and its objective is to strengthen and expand the Israeli technological infrastructure. It has a five-year budget of Euro 160 million. MAGNET encourages the early dissemination of new technologies in Israel (by subsidizing advanced technology users' clubs) and the generation of technologies that are crucial to the competitiveness of Israeli industry. R&D has to be carried out by a consortium of at least two companies and one academic institute. Partners have considerable programmatic, organizational and operational freedom. Each party is permitted to use the technology for the development of its own products, but the selling of a licence requires the approval of the partners and the OCS. Kaufmann and Yinnon (1996) have some doubts on the effectiveness of the MAGNET programme, although they consider the establishment of MAGNET to be one of the most important policy steps taken by the Chief Scientist in the last ten years. They note that the MAGNET programme is still rather unselective, supporting a large range of incoherent projects. This causes a loss of critical mass. Furthermore, communication to potential users of the programme is still not optimal and good organizational embedding is lacking. According to Kaufmann and Yinnon, an organization is needed that combines neutrality with technological knowledge in order to be in a good position to create consortia. It must help firms to identify reliable partners, provide legal support and ensure the effective operation of the consortium.

Finally, Israeli policy seems to be particularly successful in venture capital. In 1993, the government started stimulating venture capital through the establishment of Yozma Venture Capital. With a budget of Euro 80 million, Yozma has invested, along with foreign venture capital investors, in nine local venture capital funds. Apparently, Yozma acted as a catalyst to attract foreign funds, since the original government funds were multiplied by significant private (foreign) funds. Yozma's task is now completed, and the private partners of the nine local funds can buy the government out. These funds will now have to raise new money on the capital market (without government support). However, although these funds have already sold companies on NASDAQ or to larger companies, they do not show the investment returns that private investors expect (*The Financial Times*, 3 May 1996).

Conclusion

According to Teubal (1993), for years the Israeli civilian R&D capacity has been crowded out by military R&D that was needed to maintain a high deterrent capacity. Partly forced by the need to integrate many highly educated immigrants from the former Soviet Union into Israeli society, the government started to change its science and technology policy. Financial instruments became more selective, supporting a limited number of technologies, and co-ordination was improved. The Office of the Chief Scientist of the Ministry of Industry and Trade is the central co-ordinator. However, Kaufmann and Yinnon (1996) indicate that still much has to change, especially with regard to the functioning of research and intermediary organizations.

7.3   JAPAN

Introduction

The greatest difference between Japan and the case study countries lies in the perception of the role government plays in science and technology. The image of Japan as a closed, centrally or at least collusively managed economy, the 'Japan Inc.' view, has a strong hold on minds. Associated to this view is the perception that the power of the Japanese government extends far beyond its financial involvement, which is rather low as a percentage of GDP compared to other industrialized countries. The ministry most often referred to when illustrating the fierce hold of the government on the economy is the

Ministry for International Trade and Industry (MITI). MITI is also the most important ministry in matters relating to science and technology. However, the traditional view of using the 'Japan Inc.' hypothesis to explain Japan's postwar economic success is disputed by a growing number of academics. Detailed research has shown that the structure and functioning of Japanese (manufacturing) industry is quite different than commonly thought, and the role of the government, direct and indirect, smaller.

By the beginning of the 1990s there came an end to the Japanese success story and the country experienced the consequences of the 'bubble economy' of the late 1980s. Since then, growth has been small or even negative, which is an unusual experience for most Japanese. Efforts to kick-start the economy using Keynesian public expenditure type programmes have not succeeded, at least not in the way that was hoped. These programmes, for their part, have resulted in large budget deficits leading to interest payments on debt equalling some 20 per cent of total budget expenditure.

At the time the economy started to flag, the political situation in Japan started to change also. Over a period of many years the Japanese political system, which was pressed forward by the Americans after the Japanese defeat in the Second World War, proved to be very stable. The Liberal Democratic Party (LDP) held a comfortable majority for decades. At the beginning of the 1990s, after a number of bribery scandals and internal conflicts, several factions within the LDP split off. New parties emerged (mostly founded by ex-LDP members) and took over power, sometimes with the traditional opposition party, the Social Democrats. After several unstable governments, the LDP took over again in the middle of the 1990s. However, none of the governments succeeded in the fight against the recession. This fact is sometimes attributed to the conservative and powerful Japanese bureaucracy.

In 1993 Japan's gross R&D expenditures equalled 2.93 per cent of GDP. This is the second-highest (after Sweden) of all OECD countries. The level of Japanese investments in science and technology is taken as a target in several other countries, including Germany and France. However, the coverage ratio (the ratio of receipts for exported technology over payments for imported technology) of the Japanese technology balance of payments has become larger than one only recently.

Structures

In Japan the budget is presented to the Parliament usually in January. This is preceded by detailed negotiations between the Ministry of Finance (MoF) and the spending ministries. The spending ministries have to present detailed plans of revenues and expenditures; it seems that a frame-budget comparable to that of the other case study countries is not in operation in Japan. After the budget is set, each ministry is of course responsible for its own budget. The MoF monitors ministerial spending closely (discussions are held four times a year). This is underlined by the fact that the MoF bears overall responsibility for the implementation of the budget.

The most important ministries in matters of science and technology are MITI and the Ministry of Education (MoE). Most of government science and technology spending is channelled via these two ministries. Other spending ministries also house government research institutions such as the National Space Development Agency and the Marine Science and Technology Centre. Education in Japan is a shared responsibility between national and local governments. The latter run the schools, while the former is by law obliged to provide half the teachers' salaries in public schools. A particular feature of the Japanese education structure is the high share of engineering graduates as opposed to graduates in the humanities or 'pure' sciences.

Philosophies

To the outsider, the main philosophies of Japan regarding science and technology seem to be characterised by opportunism and mission orientation. The former hints at the remarkable capability of Japanese industry to adapt information from abroad to new products and production methods. The latter is a description of Japanese science and technology policy. Japanese political culture seems to favour bargaining. As one of our interviewees said, the philosophy of government science and technology in Japan is no different from the West: it is mainly a question of handing out perks. This is evident in that, although small in total, a large share of government science and technology spending has gone to large-scale projects such as (nuclear) energy, telecommunications, computers and space. In other fields, Japanese policy is actually remarkably hands-off and there is convincing evidence that competition between Japanese firms is high. In 1993 the government funded only 21 per cent of total R&D, which is the lowest percentage of all OECD countries. Furthermore, the government performed only 10 per cent of R&D activities, which is also very low. There is even a decreasing trend in government intervention in all sectors of the economy. This is quite unlike the traditional view of 'Japan Inc.'. However, even in the 1960s, the heyday of public policy intervention, the influence of government (as measured by direct intervention) was limited, as recent research has shown (Miwa, 1996). Another commonly held view of the Japanese system is that it is co-operative. As evidence on the science and technology side, reference is made to the joint research ventures fostered by MITI. As we shall discuss in the next section, these have turned out not to be a general success and even the degree of co-operation has been disputed.

Policies

One policy tool that has attracted considerable attention both from academics and western politicians is research joint ventures. These are based on an Act dating to 1961. Firms were given the opportunity to establish research associations and to apply for government funding. In the period 1961–87, 87 such associations were established. As Odagiri and Goto (1993) note, most of government support for industrial science and technology was channelled via these associations. They have also been a channel to support the mission-orientated programmes, such as semiconductors and computers. The success of this policy, channelling targeted funds through government-supported industry-orientated organizations, has often been exaggerated. We recall that government support for industrial science and technology is low; therefore, support for specific programmes ('picking the winner') is also relatively low. Research associations are by no means universally considered to have been a success. Research by Wakasugi and Goto (1985) and Fujishiro (1988) found that the associations produce fewer patents for a given amount of science and technology expenditure than industry in general. In interviews people indicated that research associations are now seen (at least in academic circles) as an unsuccessful attempt by the government to 'beat the market'. The most favourable opinion about them was that they allow the administration to minimize monitoring efforts. The success of mission-orientated programmes, for example on nuclear energy, have been questioned as well. Whatever the truth may be, since the fast-breeder reactor started production it has proved to be a commercial and technical flop (the latest reports suggest it has consumed more energy than it will ever produce).

Of course informal ways and non-financial policies to influence industry science and technology decisions surely exist, just as in other countries. However, these would have to be enormously effective to counterbalance the modest financial tools in place. Certainly, earlier (also western) research emphasized the importance of non-financial measures, for example non-tariff trade barriers and policies regarding patent applications. Nester (1991) cites several cases where foreign patent applications have been delayed, with the result that Japanese firms have filed for very similar patents. He also maintains (and this was confirmed in our interviews) that the Japanese patent system is difficult for foreigners because, for example, the applications have to be in Japanese. Another example of non-monetary policy concerns the existence of specific Japanese standards, as has been the case with high-definition TV and mobile phones. In both cases, the standards have no doubt prevented foreign firms from entering the Japanese market. However, at the same time this approach made Japanese firms captive to standards that were out-dated and not in use anywhere else. So instead of improving competitiveness of Japanese firms, the policy has probably produced the opposite result.

Conclusions

The myth about the Japanese economy and government policies being somehow clearly different from those in western countries seems to be unravelling. This applies to science and technology policies as well. Just looking at the figures, the striking things are the low share of government expenditure and the high share of education within government expenditure. A considerable share of public science and technology funding is directed towards large-scale projects, some of which have proved a success, some a failure. Anyway, the Japanese government has not demonstrated a capacity to beat the market, and it seems that the target areas chosen reflect as much political considerations as economic ones.

If Japan does anything differently with respect to science and technology matters, this is on the non-monetary side, reflecting (possibly) cultural differences and the like. There is plenty of anecdotal evidence that various government procedures are organized in a way that favours Japanese firms over foreign firms. This is not a feature peculiar to Japan, although the common opinion (although without proof) seems to be that the Japanese are more efficient in this respect than other nations.

7.4   SINGAPORE

Introduction

Singapore is a city-state just south of the peninsular Malacca with about 3 million inhabitants. It became fully independent (from Malaysia) in 1965 after racial conflicts between the Chinese (a majority in Singapore) and the Malay. In its early years of independence Singapore faced daunting economic challenges. It had no natural resources, the population was growing rapidly and unemployment was high at about 10 per cent. Singapore's economy was highly dependent on entrepôt trade and on the provision of services to British military bases. Furthermore, it had a small manufacturing base, little industrial know-how and hardly any domestic entrepreneurial capital. In the years following independence, foreign investors were attracted to Singapore to develop the manufacturing and financial sectors. In many ways the Singapore government tried to improve the climate for investment. It enacted the Employment Act and the Industrial Relations Act which laid down standards of employment and dispute resolution procedures. It invested in key infrastructure and established new companies in areas where the private sector lacked capital or expertise. Whether or not this can be attributed to government policy, the Singapore economy grew at an average rate of 10 per cent annually during the period 1965–80. Unemployment fell steadily until it reached 3 per cent in 1980. A strong manufacturing sector developed, accounting for 28 per cent of GDP in 1980 (26.7 per cent in 1996), up from 15 per cent in 1965.

In 1980 Singapore faced a new challenge of restructuring the economy towards higher value added activities (both in industry and in services) while facing at the same time a very tight labour market. Consequently, a renewed emphasis on education and training emerged and investment promotion policies became more selective. The government encouraged investment in high-tech industry or itself participated in new companies. By 1990 Singapore's economy had become mature, with income levels equalling those of North American and European countries. However, other problems emerged. Companies and individuals preferred to invest in booming real-estate markets instead of in uncertain R&D activities. The challenge for Singapore is now to consolidate its position in technology.

As described above, the government was strongly involved in the development of the Singapore economy. Like some other Asian governments, the Singapore government can be defined as an 'enlightened authoritarian'. It combines economic liberalism and authoritarian government, and considers it the ideal way to bring welfare to the entire population. Confucianism is its cultural root: family, savings and hard work are central concerns in ones personal life. Tight government control is used in Singapore as a means to prevent unemployment, racial conflicts and crime. Government power is indirectly exerted through statutory boards and government linked companies (GLCs). By 1991 there were 419 of these GLCs and statutory boards. Today, even the Singapore government must admit that there are some problems within the society. There are signs that Singapore lacks the ability to create its own technological base and that an over-protective authoritarian government oppresses the necessary creativity. Politics are dominated by the People's Action Party (PAP), which is led by President Ong Teng Cheong, and any attempts to democratize Singapore society have been effectively thwarted.

Structures

Within this present policy framework the National Science and Technology Council, chaired by the Minister for Trade and Industry, is responsible for the overall co-ordination of science and technology policies. Three funding institutions (the National Science and Technology Board (NSTB), the Ministry of Education and the Ministry of Health) work together to bring about developments in various areas of science and technology. Cluster agencies (PSB,

EDB (see below) and others) are primarily responsible for the investment promotion and economic performance of a particular sector. In the end, NSTB, the relevant cluster agency and the relevant cluster research institute will be collectively responsible for the technological upgrading and performance of a particular industry.

Established in 1991 as a statutory body, the objectives of NSTB are to develop Singapore into a centre of excellence in selected areas of science and technology and to enhance national competitiveness in industrial and service sectors. To complete its tasks NSTB creates appropriate financial assistance schemes, co-ordinates the establishment of research institutes and other science and technology facilities, and assesses science and technology manpower needs. NSTB also engages in exchange and joint science and technology programmes with international organizations and other countries. The Singapore Productivity and Standards Board (PSB) promotes effective technology applications in industries in order to increase productivity. PSB collaborates closely with universities and research institutions, both locally and abroad, to achieve its objectives. Its target groups are multinational corporations, promising SMEs and industry-wide cluster groups. The Singapore Economic Development Board (EDB) is charged with the responsibility of drawing in foreign investments and technology, creating jobs and promoting economic development through industrialization. Together with the NSTB, EDB draws up plans and strategies for building up Singapore as a R&D hub in the Asia-Pacific region.

R&D activities are promoted in two universities (the National University of Singapore and the National Technological University) and four polytechnics. The universities are responsible for producing well-trained graduates. Traditionally, R&D in the universities tend to be centred around individuals. The universities have recently started to restructure research by building groups of researchers around laboratories to create a greater impact. Included in the technological infrastructure is a network of 13 research institutes and centres funded by the NSTB, which is also responsible for their strategic development. The research institutes and centres train R&D manpower for industry, develop pre-competitive technologies and provide R&D infrastructural services and transfer technology to industry. The research institutes are positioned to collaborate with a specific industry or group of industries where generic technologies are developed. Most of the research institutes, national technology programmes and industry consortia are targeted at existing industry clusters. However, a few are proactively fostering the growth of new high-tech industry clusters. In the future, research institutes will receive greater operational autonomy, but at the same time a Research Institute Management Council will be formed to provide strategic direction and policy guidance to the institutes. Between research institutes and industry a 'buyer–

seller' relationship will be established to achieve greater orientation towards outputs which serve the needs of industry.

Philosophies

During 1978–94 R&D expenditures have increased significantly. Over that period R&D was always more than 50 per cent financed by the private sector. However, government expenditures on R&D have increased faster over the years. In 1994, R&D expenditures equalled 1.12 per cent of GDP, which is significantly lower than in other newly industrializing countries such as Taiwan and South Korea. Since the Singapore government generally compares itself with these countries, this is a major concern of the government. Furthermore, with regard to the number of research scientists and engineers per 10 000 inhabitants, Singapore (42 in 1994) is behind South Korea and Taiwan (57 and 61 respectively). In 1994 engineering accounted for more than half the total R&D expenditures, while computer sciences accounted for nearly a quarter. Natural sciences and biomedical sciences accounted for 14 per cent and 4 per cent respectively.

For Singapore to be trailing is nothing new. Shortly after the foundation of Singapore as an independent state little was done to promote technology development. Technology was mainly transferred by multinationals with production plants located in Singapore. The research remained in the multinationals' home countries. Gradually Singapore benefited from 'learning by doing', and technologically more sophisticated production processes were located in the small city-state. Only by the end of the 1980s did the government become more actively involved in R&D. It participated in high-tech industry and introduced tax incentives favouring the adoption of new technologies. In 1991 the Singapore government presented its vision on science and technology in the National Technology Plan. The Plan aimed at the growth of new high-value-added industries. It suggested the development of the technology infrastructure, the creation of incentives for domestic and foreign industry to invest in R&D and the development of human resources to support R&D. Targets were set for R&D expenditures equal to 2 per cent of GDP and achieving a ratio of 40 scientists and engineers for every 10 000 workers. Moreover, government financing of R&D was never to exceed 50 per cent of total R&D expenditures. The first target was not met (the ratio was 1.12 per cent in 1995) and is currently set at 2 per cent for the year 2005. Both of the other targets had already been achieved by the start of the Plan or soon afterwards.

In 1996 a new plan was published. Singapore's aim for the next five years was to have an annual GDP growth rate of 6–8 per cent. According to the government, Singapore has to become an international business hub and a centre of high-tech industries. Bottlenecks that hindered the development of

Singapore science and technology capabilities will be removed, most prominently the lack of highly educated manpower. Other bottlenecks are the rising cost level and the commercialization of technologies. The government intends to encourage companies and individuals to undertake R&D activities. This means more resources in the form of R&D grants. The Singapore government seems to be very generous in this respect. It will finance up to 50 per cent of the costs of near market projects, and for other types of research this percentage can be even higher. To solve the human resource bottleneck the government wants to attract talented foreign researchers. However, living conditions are considered to be unattractive on the densely populated island ruled by an authoritarian overprotective government. Young Singaporeans are given various incentives to choose a career in research. However, other professions (for example in the financial and real-estate markets) have been at least financially more attractive. Another spearhead of Singapore science and technology policy is the improvement of the availability of knowledge. Overseas co-operative agreements for technology co-development are established, and research centres can be set up with Singapore money under certain conditions regarding knowledge transfer. An agency will be established (probably under the wings of the NSTB) 'to undertake the role of technology sourcing, and to create a national repository of new and emerging foreign technologies'.

Not included in the recent Science and Technology Plan but nevertheless relevant to technological development is direct government participation in high-tech activities. This takes place when capital-rich government-linked corporations (GLCs) and statutory boards invest directly in technology acquisition and development as equity owners of foreign high-tech companies or partners in joint venture strategic alliances with foreign multinationals. The most important GLC to engage in strategic technology investments at home and abroad is Singapore Technology Holdings (STH), which has invested in venture capital high-tech start-ups in the US, particularly in Silicon Valley, to secure the transfer of technology and manufacturing capability to Singapore.

Conclusion

The Singapore government is ambitious with regard to the future. In order to be able to continue the high growth of the recent past, the government aims to bring R&D expenditures up to 2 per cent of GDP in 2005. Furthermore, it intends to take a range of measures to make Singapore a high-tech country, both in industry and in services. However, plans have not proved completely successful in the past. The objectives of the National Technology Plan 1991 were hardly or not at all realized. Furthermore, the economic crisis in other Asian countries might hit the open Singapore economy severely.

7.5   TAIWAN

Introduction

Modern Taiwan was established when the nationalist government of China moved its seat of government to the island of Taiwan (at that time known as Formosa), leaving the mainland to their communist enemies to establish the People's Republic of China. The relationship between the two states is still rather hostile. However, it is important to differentiate between rhetoric and facts: the fact is that many Taiwanese consider Taiwan's claim to the mainland to be unrealistic. However, dropping that claim implies proclaiming independence, which would never be accepted by the government of the People's Republic, which more or less tolerated Taiwan as a disloyal province (that is, part of China). On the other hand, Taiwan is one of the biggest investors in China's booming economy. China's 'one nation, two systems' solution to the conflict is met with scepticism in Taiwan, given the resistance of the Chinese government to increasing political freedom in their own territory. In fear of economic retaliation by communist China, most countries in the world do not officially recognize Taiwan as an independent country but maintain informal relationships with its government. Furthermore, as a consequence of the increasing economic power of communist China and the withdrawal of former colonialist powers from Macao and Hong Kong (returning these cities to the Chinese government) Taiwan's international political position is slowly weakening. However, over the years Taiwan has always been backed by the US. Its uncomfortable international position forces Taiwan to depend heavily on its economic power, and for that reason economic and technological development are very important.

Rapid economic growth (about 8.6 per cent annually from 1952 to 1995) and a comparatively equal distribution of wealth gave the 21 million inhabitants of Taiwan one of the highest standards of living in Asia. The foundation of the present economic well-being of Taiwan, which has virtually no natural resources, was laid by land reforms in the 1950s and the attraction of foreign investors looking for cheap labour in the 1970s. At the end of the 1980s, after domestic labour costs had significantly increased, Taiwan ceased to be a cheap labour country. Taiwanese companies even started to invest in countries with cheap labour such as Indonesia and mainland China. Today, Taiwan's wealth comes mainly from the export of industrial products such as machinery, electrical equipment, transport equipment, textiles, base metals, manufactures, plastic articles and toys. However, the government plans to shift production towards high-tech, high-quality products. Unlike the South Korean economy (with which Taiwan is often compared) the Taiwanese economy is not very concentrated. Large Taiwanese conglomerates do exist (*guanqixiye*), but their influence on the Taiwanese economy is significantly less than that of the Korean *chaebols* on the South Korean economy. The Taiwanese economy is dominated by SMEs with fewer than 300 employees, accounting for 60 per cent of Taiwanese exports. With regard to R&D this is considered to be a weakness of the Taiwanese economy, as is the shortage of sufficiently qualified and experienced personnel and the lack of foreign and domestic market information for Taiwanese producers.

Formally Taiwan has a democratic political system. However, it has been ruled by the Nationalist Party ever since 1949. It is only in recent years that political parties have emerged which have threatened the dominant position of the Nationalists and democracy has become more or less established. The Taiwanese government has a firm grip on the economy through companies owned by the state or the Nationalist Party (including banks and insurance companies).

Structures

Each ministry has its own Office of Science and Technology Advisers responsible for hiring science and technology advisers and serving as a science and technology consultant to its minister. The Ministry of Economic Affairs (MoEA) is responsible for national economic administration and economic construction, with a scope of functions encompassing industry, commerce, trade, energy and mining. The Industrial Development Bureau is the staff unit that is responsible for technology policy. MoEA is also responsible for several public research institutes and publicly owned companies that account for nearly 10 per cent of national R&D expenditures. These companies are also strongly involved in research. The China Petroleum Corporation, BES Engineering Corporation, China Steel Corporation and the Taiwan Power Company own some of the largest research institutes. The National Science Council (NSC) is charged specifically with promoting, planning and co-ordinating basic research and science education. In addition, the NSC is responsible for developing science-based industrial parks. Members of the NSC are the heads of those government institutes which carry out science or technology projects, the ministers in charge of reviewing scientific or technological development, the president of Academia Sinica, a high-ranking civil servant and noted scientists.

For each area of research there are special funds and specialized public research institutes. The numerous private research institutes in Taiwan focus mainly on experimental development and the 'commercialization' of R&D results. Some of the more active research institutes in Taiwan include the Industrial Technology Research Institute, the Institute for Information Industry and the Food Industry Research Institute. The Academia Sinica is the leading research institution of Taiwan. Its two basic missions are to conduct scientific research in its own institutes, and to direct, co-ordinate and promote scientific research in other institutes and universities throughout Taiwan. Although it is an integral unit of the government under the direct supervision of the Office of the President, the Academia Sinica is practically independent. The Industrial Technology Research Institute (ITRI) is Taiwan's largest research institute (it has about 5800 employees, 70 per cent of them engineers and researchers). ITRI has been charged with special tasks regarding industry, such as the upgrading of existing industries and the promotion of new ones. Another task is to keep Taiwan in touch with new technological developments worldwide. ITRI has been subject to ambiguous criticism (Wang, 1995). On the one hand, firms claim that ITRI develops technologies that do not match their needs; on the other, ITRI is criticized for engaging too much in short-term technological development projects. In addition, it is stated that ITRI develops advanced technologies for many industries without careful selection, resulting in ineffective technology development. Finally, ITRI is criticized for relying too much on expensive internal R&D instead of importing technologies from other countries. A solution that is being considered is to redefine the tasks of various research institutions (ibid.). ITRI will engage in applied research, half of which will be funded by the private sector (for many years ITRI has been mainly financed by the government). The universities will, more than before, be charged with basic research. Whatever the exact plans, firms will be charged with more responsibility regarding near market R&D.

Philosophies

Today the Taiwanese government funds about 50 per cent of the national R&D effort which is very much concentrated in engineering R&D. The first step towards modern science and technology policy was made in 1979 when the Science and Technology Development Programme was set up. As a result of this programme, changes were made in the governmental administrative structure to strengthen inter-ministerial co-operation in science and technology projects. The programme encouraged venture capital for high-tech industries, disseminated technology to accelerate the upgrading of industries, and attracted overseas Chinese scientific and technical manpower to start up businesses in Taiwan. In the 1980s, as Taiwanese industry became increasingly threatened by competitors from neighbouring East Asian countries, it became clear that Taiwan needed a national innovation system that provided the capacity not only to receive and to absorb the (imported) technology, but also to adapt, diffuse and (ultimately) improve it. Furthermore, the Taiwanese government looked for public–private partnerships to guide and to co-ordinate activities which ensure that high-tech industry would remain at the cutting edge of technological developments (Matthews, 1995).

Since the middle of the 1980s, the Taiwanese government has striven to make Taiwan an 'Island of Technology'. New and extended research institutes are intended to contribute greatly to Taiwan's advancement in academic research, high-tech industrial development and manpower development. Companies, state-owned as well as private, are encouraged to maximize their investment in R&D activities in particular promising sectors. Co-operative research is encouraged to overcome the problems that SMEs, which are very dominant in the Taiwanese economy, experience in carrying out risky R&D. Co-operation with multinationals and foreign research institutes is stimulated. To ensure the supply of high quality labour, policies have been put in place to encourage Taiwanese students studying abroad to return to their home country. Also, to encourage foreign investors, patent protection has been improved.

The Taiwanese government is criticized for the fact that an efficient and effective science and technology policy is severely hampered by the inability to overview and direct the institutional system supporting it. There are many responsible authorities and rules and frequent changes of rules and responsibilities. There are numerous cases of conflict among ministries that are evidenced in many policy priorities in different parts of the government. The directives are often vague and uninstructive, which limits the legal security of potential investors. The National Science Council's performance over the years has been severely hampered by a chronic shortage of financial resources, a lack of political influence and a general lack of statistical data. The NSC must often compete with more powerful organizations for influence and resources. Moreover, tax incentives, credit control and the use of public enterprises might have been three effective policies to foster targeted industries. However, as evidenced by Schive (1995), the Taiwanese government applied all these three policies with a high degree of self-restraint. Finally, there seems to be a need for evaluation procedures for projects and policies.

Policies

The Technical Manpower Cultivation and Recruiting Programme was set up to upgrade scientific and technical manpower at universities and research institutes. Several other plans were developed in order to utilize fully Taiwan's limited scientific and technical resources, to develop industrial technology and to strengthen academic research. Manpower development, however, has not been very successful. Although MoEA has universities and non-profit organizations to handle research manpower training, the efforts are not very encouraging and the lack of R&D staff remains a big concern for

Taiwan. The major problem is that Taiwan's educational system focuses on educating teachers and professors, while the training of technical research manpower is neglected. Because there is no connection between education and research, a shortage of research manpower has resulted. Several adjustments, such as the co-ordination of educational planning with research manpower demands, encouragement of co-operative research and efforts to increase the education of researchers, are considered necessary.

MoEA supports the development of new products in high-tech industries (such as communications, aerospace, advanced materials, semiconductors and so on) by funding 50 per cent of the development costs. The conditions are straightforward: the technology necessary for the development of the product must exceed the existing level of technical expertise in the domestic industry, and the product must have a high market potential and the capability to stimulate the development of related industries (MoEA, 1995b). Looked at in this way, Taiwanese science and technology policy does not differ from western policies. However, the Development of Critical Components Programme (DCCP) reveals some typical features. Essentially, it was established to reduce Taiwan's trade imbalance with Japan and to promote the industrial upgrading of Taiwan. For products or technologies to be eligible for the DCCP programme, value added and development potential must be high. However, the new technology must replace imported goods in large quantities. This objective refers to old-fashioned import substitution policies.

7.6   UNITED STATES OF AMERICA

Introduction

For a large part of this century the US succeeded in being the unchallenged world leader in a wide range of technological areas. However, over the last decades this US technological lead has gradually shrunk. A number of indicators suggest that since 1973 US living standards have stagnated or declined in real terms, aggregate productivity growth has remained low on average and exports have tended to lag behind imports. Other countries have caught up with or even surpassed the US in R&D investment, productivity and income. The reduced performance of the US in many dimensions (except unemployment, which is relatively low) is either seen as, or attributed by many to, a decline in international competitiveness resulting from a failure to maintain its technological lead. Various explanations for this loss of technological lead have been advanced. Some concern issues like the organisation and financing of R&D or the exploitation of results. Among the more contentious ones are those suggesting that the general belief in the efficiency of free and competitive markets has been harmful to technological development. For example, the emphasis upon anti-trust issues gave rise to concerns that anti-trust legislation has slowed the innovation process. Not only can anti-trust regulation hamper R&D collaboration among firms; it can also prevent innovators reaping full profits. Over recent years anti-trust regulations regarding R&D collaboration have been relaxed.

In 1996 the national R&D effort amounted to $184.3 billion. Of this amount 62 per cent was spent by industry, while the government spent about 34 per cent. Industry is not only the largest spender on R&D, but is, with 73 per cent, also the largest performer. We conclude that there is a significant government funding of industry-performed R&D, mainly in defence. Government expenditures on industrial development have been very small, reflecting the reluctance of the federal government to enter into civilian, and therefore more commercial, R&D. Over the period 1990–7 total R&D expenditures have increased but government R&D expenditures have decreased. This can be attributed mainly to decreasing defence R&D. US defence-related R&D expenditures reached their highest level at the end of the Reagan administration in 1987–8. At that time government defence R&D expenditure was more than twice as large as non-defence R&D.

Structures

The political system of the US is quite unlike that of most European countries. In countries such as the UK and Finland the executive government is elected by parliament and consists of one or more parties that together have a majority in parliament. In the US the President is elected directly by the citizens, as is the Congress. Therefore, it may happen that a US President is elected who has no majority support in Congress. For many years during the last two decades this has been the case. The USA has a system of checks and balances, with the Congress having the potential to exert considerable control over the actions of the President. In those European countries that have coalition governments, the debate between the coalition partners within the cabinet is the forum where compromises are reached. In the US, finding workable compromises is rather a matter between the President and his administration on the one hand and Congress on the other. So although the general public expects the President to lead and give directions (and holds him accountable when he does not), his power is in fact rather limited if the Congress is not willing to co-operate. The major control of the President comes from his appointment powers. He can draw on a wide range of people to be Cabinet members; some of their nominations have to be confirmed by Congress.

Under the President is the federal government. This contains a remarkably diverse collection of organizational types: executive departments, independent agencies, commissions and boards. Most federal activities and responsibilities are conducted through the executive departments. These departments supervise a wide range of other agencies and bureaux that report formally to them. In many cases the agencies collected under a departmental umbrella have conflicting or competing missions. Moreover, what each department does is a legacy of past political battles and not necessarily rational. However, by no means all federal government activities are organized in or through executive departments. Independent agencies and commissions, for example the Environmental Protection Agency, administer many activities. To complicate matters further, departments and agencies face demands for political responsiveness from the Congress as well. Unlike many parliaments in the world, the Congress is a significant player from the perspective of executive departments and agencies. The Congress legislates any significant change in the structure of federal organizations that is proposed by the President. It has an effective veto over any reorganization. Departments and agencies have to justify their budgets during the annual budget setting process as well as obtain periodic authorization for their programmes. Appointments of top officials need the approval of Congress and are obliged to enforce the law in line with congressional intent. In order to fulfil these tasks the Congress has an extensive staff and a very complex structure of committees and sub-committees, whose responsibilities often overlap. These committees have considerable influence on the content and funding of government policies. It has been argued that such close congressional oversight through its committees can result in slow, risk-averse and non-entrepreneurial decision making. Assertive decision making is sometimes hampered by professionally organized and often wealthy interest groups.

The Department of Defense, the Department of Health and Human Services, the Department of Energy, the National Aeronautics and Space Administration (NASA) and the National Science Foundation (which is responsible only for basic research and science education not tied to any particular federal mission) spend most of the federal R&D budget. However, within these broad departments there are lower-level agencies that matter, and responsibilities for spending decisions are thus widely spread.

Philosophies

The fundamental charter for American postwar science policy and its general philosophy was the Bush Report (Bush *et al.*, 1960). This report recommended the use of public funds to support basic research in colleges and universities and to 'foster the development of scientific talent in our youth'. Research was to be largely supported by contracts and grants with universities and research institutes as well as private firms, leaving internal control of policy, personnel and the method and scope of research to the institutions themselves. Science was thus to be given wide self-governance and autonomy. Not all suggestions of the Bush Report were realized. Instead of a single agency responsible for all extramural research sponsored by the federal government, there evolved a pluralistic support system with many organizations responsible for the support of research. Furthermore, the Bush Report recommended the principle of awarding research and development contracts to the most qualified organization irrespective of geographical or other non-scientific considerations. Brooks (1986) notes this to be particularly in conflict with the apparent desire of members of Congress to please their local constituencies.

It is fair to state that for most of the postwar period the federal government considered its appropriate role as largely restricted to defence research and basic (university) research. The role of the federal government in the US innovation system was not linked to any economic strategy, being largely motivated by national security concerns (Mowery and Rosenberg, 1993, p. 62). Little, if any, strategic planning underpinned public intervention in the US innovation system and policy makers devoted minimal attention to its domestic economic payoff. Unlike in other countries (such as Technology Foresight in the UK) there have been no explicit attempts to direct government support for R&D towards those areas that might have yielded the greatest national benefit. For a long time, US science and technology policy was mission-orientated in the extreme (Ergas, 1987). Civilian R&D was not considered to be an important government task. However, by 1988 things started to change as it became clear that other countries started to match or improve upon US performance. At the same time government intervention in the technology process of the East Asian tiger economies seemed to be successful (see Krugman, 1996). The US government started to get engaged in civilian R&D on a small scale. Shortly after taking up the Presidency (22 February 1993), President Clinton issued a document entitled *Technology for America's Economic Growth: A New Direction to Build Economic Strength* as a technology policy manifesto. This policy statement clarifies where concerns with US technological performance lay:

> Investing in technology is investing in America's future: a growing economy with more high skill high wage jobs for American workers; a cleaner environment where energy efficiency increases profits and reduces pollution; a stronger, more competitive private sector able to retain USA leadership in world markets; an educational system where every student is challenged; and an inspired scientific and technological research community focused on ensuring not just our national security but our very quality of life.

Remarkable, however, is the statement about the role of the government regarding R&D in the private sector that contrasts to the general American belief in the efficiency of the free market. The document states:

> The traditional federal role in technology development has been limited to support of basic science and mission oriented research in the Defense Department, NASA and other agencies. This strategy was appropriate for a previous generation but not for today's profound challenges. We cannot rely on the serendipitous application of defense technology to the private sector. We must aim directly at these new challenges and focus our efforts on the new opportunities before us, recognizing that government can play a key role helping private firms develop and profit from innovation.

However, budget deficits virtually ruled out new federal spending and made cuts in existing expenditure a high priority. Any newly found belief in the benefits of intervention has thus been tempered by budget issues.

Brooks (1986) attempts to assess attitudes to the federal role. Although this view is now 12 years old and there have been two Presidents in that time, a number of things have changed only marginally. In the US there is a general consensus on the arguments in favour of government support for R&D when there is market failure, when public goods are involved and when the government is the ultimate user of results. Defence R&D is a case in point, but so are environmental protection and health and safety issues.

Brooks then points at two areas where according to him the real controversies about government intervention in the innovation process exist: the 'key industry' and the 'generic applied research' issues. The proponents of federal support in these areas argue that there are certain industries and technologies which are so fundamental to the future development of an economy that federal support is required to ensure future prosperity (especially if overseas governments are giving support to these areas). The opponents argue that such key technologies would be recognized by the private sector and, if potentially profitable, that sector would undertake the necessary investment. Currently, more than a decade after Brooks wrote his analysis, this discussion seems to have been won, in the US like in most other countries, by those that argue against attempts by government to identify key technologies and pick winners.

Internationally, the US government continually argued for stronger intellectual property rights and tried to reduce barriers to its exports or perceived unfair subsidies to industries in competing nations. The US has also been active in placing barriers on high-technology exports from the US when it was considered that those exports might have defence implications (although some perceived such restrictions to be at least partly driven by commercial rather than defence objectives). The pressure in the US to create 'level playing fields' for international trade appear to have been particularly strong and have grown stronger as the US technological lead has been reduced (Ostry and Nelson, 1995).

Policies

Federal science and technology policy is largely directed at specific public objectives, such as defence, environmental protection, or the general advancement of knowledge. It seems to be only accidental that these policies encourage private R&D as well. On the other hand, defence R&D is sometimes supposed to have significant spin-offs to the civilian sector, and therefore government support for defence R&D is again justified. Whether this spin-off has actually been forthcoming is not totally clear. One may think of examples like the aerospace industry where there have been spin-offs, but the literature is by no means definitive in arguing that the civil spin-offs from defence research have been substantial (Stoneman, 1987). In any case, to drive a civil technology policy via defence R&D is a rather roundabout route to the objective.

Despite its obvious reluctance, the federal government has two important instruments to stimulate private R&D in an indirect and market-like way: the tax credit system and the Small Business Development Act. The tax credit system enables firms to offset a rise in R&D expenditures against tax. The tax credit system can be seen as the major route by which the US supports industrial R&D expenditure. It is fair to state, however, that there has been considerable argument in the literature as to whether the policy has in fact increased R&D spending in the US. What is obvious is that the tax credit system is very wide based and contains no strategic dimension. It is still left to the market to determine where R&D spending should be directed. In July 1982 Congress enacted the Small Business Development Act, which mandated all federal agencies spending more than $100 million annually on external research to set aside 1.25 per cent of their budget (2.5 per cent from 1997) for awards (grants or contracts) to small businesses. This now involves sums approaching $1 billion per annum. According to Lerner (1996), the Small Business Development Act has had a considerable impact upon high-tech small firms.

Conclusion

US government involvement in the national innovation system is extensive but is concentrated on defence, high-tech and basic science. US science and technology policy is not driven by concerns about industrial technological development. Although there is concern over the industrial technological performance of the economy, this has not led to large increases in direct spending to address such concerns, neither is there any strategic underpinning to the spending that does take place. Despite the statements of Bill Clinton when he became President, there is still a reliance on the serendipitous application of such R&D support to the private sector. There is little emphasis, if any, on diffusion issues in US science and technology policy, although it is generally believed (but never proved) that, for example, defence R&D has large spin-offs to civilian industry.

Changing US science and technology policy is practically impossible due to the country's political system. The weak party system, the large influence of interest groups, the eventual lack of political support in Congress for the President and the extensive power of Congress over federal bureaucracy ensure that: (a) new initiatives can come from many different origins; and (b) there is close monitoring of the performance of the executive by the legislature through the Congressional Committee system. The widely spread nature of responsibilities between the President, individual Secretaries of State, separate agencies and Congress also makes the development and institution of strategic approaches to science and technology most difficult.

REFERENCES

Israel

Jerusalem Institute for Israel Studies (JIIS) 1993, *Newsletter*, May.
Kaufmann, D. and T. Yinnon, 1996, 'Generic R&D collaboration between firms: the Israeli experience', in M. Teubal (ed.), *Technological Infrastructure Policy*, Kluwer.
Ministry of Science and Arts (MSA), 1994a, *Science in Israel*, Nr. 5, Jerusalem: Ministry of Science and Arts.
——, 1994b, *Review of the Minister, Mrs Shulamit Aloni in the Knesset: Activities of the Ministry*, Jerusalem: Ministry of Science and Arts.
Ministry of Industry and Trade (MIT), Office of the Chief Scientist, 1996a, *Government Encouragement for Industrial Research and Development in Israel*, Jerusalem: MIT.
——, 1996b, *Israel, Technologies and Innovations*, Jerusalem: MIT.
Teubal, M., 1993, 'The innovation system of Israel: description, performance and outstanding issues', in R.R. Nelson, *National Systems of Innovation*, New York: Oxford University Press.
Yinnon, A.T., 1996, 'The shift to knowledge-intensive production in the plastics-processing industry and its implications for infrastructure development: three case studies – New York State, England and Israel', *Research Policy*, 25: 163–79.

Japan

Fujishiro, N., 1988, *The Role of Joint R&D in the Computer Industry*, master's thesis, University of Tsukuba.

Miwa Yoshiro, 1996, *Firms and Industrial Organization in Japan*, New York: New York University Press.

Nester, W.R., 1991, *Japanese Industrial Targeting: The Neomercantilist Path to Economic Superpower*, New York: St. Martin's Press.

Odagiri, Hiroyuki and Akira Goto, 1993, 'The Japanese system of innovation: past, present and future', in Richard R. Nelson (ed.), *National Innovation Systems: A Comparative Analysis*, London: Oxford University Press.

Wakasugi, R. and A. Goto, 1985, 'Joint R&D and technological innovation', in Y. Okamoto and T. Wakasugi (eds), *Gijutsu Kakushin to Kigyo Kodo*, Tokyo: Tokyo University Press.

Singapore

Lim, L., 1993, '*Technology Policy and Export Development: The Case of the Electronics Industry in Singapore and Malaysia*, University of Michigan.

National Science and Technology Board (NSTB), 1996, *National Science and Technology Plan ... Towards 2000 and Beyond*, Singapore: National Science and Technology Board.

Peebles, G. and P. Wilson, 1996, *The Singapore Economy*, Cheltenham: Edward Elgar.

Tang, H.K. and K.T. Yeo, 1995, 'Technology, entrepreneurship and national development: lessons from Singapore', *International Journal of Technology Management*, 10 (7/8): 797–814.

Wong, Poh-Kam, 1993, *Small, Newly Industrializing Economies Facing Technology Globalization: A Singaporean Perspective*, Singapore: National University of Singapore.

Taiwan

Aberbach, J., D. Dollar and K. Sokoloff, 1994, *The Role of the State in Taiwan's Development*, New York: M.E. Sharpe.

Chi-Ming Hou and San Gee, 1993, 'National systems supporting technical advance in industry: the case of Taiwan', in R.R. Nelson (ed.), *National Innovation Systems: A Comparative Analysis*, Oxford/New York: Oxford University Press.

Chung-Hua Institution, 1995, *Technology Support Institutions and Policy Priorities for Industrial Development in Taiwan, R.O.C.*, Taipei: Chung-Hua Institution for Economic Research.

Fields, K.J., 1995, *Enterprise and the State in Korea and Taiwan*, Ithaca/London: Cornell University Press.

Li, Kuo-Ting, 1995, *The Evolution of Policy Behind Taiwan's Development Success*, Singapore: World Scientific Publishing Co.

Mathews, J., 1995, *High-technology Industrialization in East Asia: The Case of the Semiconductor Industry in Taiwan and Korea*, Taipei: Chung-Hua Institution for Economic Research.

Mechthold, M., 1994, *Regionalokonomische effekte der Technologieintensiven Industriebetriebe in Taiwan*, Hannover: Geographisches Institut.

Ministry of Economic Affairs (MoEA), 1994, *Strategies and Measures for the Development of the Top Ten Emerging Industries*, Taipei: Industrial Development Bureau, Ministry of Economic Affairs.

——, 1995a, *Development of Industries in Taiwan, Republic of China*, Taipei: Industrial Development Bureau, Ministry of Economic Affairs.

——, 1995b, *The Republic of China on Taiwan, Partner for Technology and Investment*, Taipei: Department of Industrial Technology, Ministry of Economic Affairs.

——, 1995c, *To Develop Taiwan into an Asia-Pacific Manufacturing Centre*, Taipei: Industrial Development Bureau, Ministry of Economic Affairs.

——, 1995d, *Technology Development 1994*, Taipei: Department of Industrial Technology, Ministry of Economic Affairs.

——, 1996, *MoEA Technology Development Programme 1996*, Taipei: Department of Industrial Technology, Ministry of Economic Affairs.

Schive, C., 1995, *Taiwan's Economic Role in East-Asia*, Significant Issues Series, Washington, DC: Centre for Strategic and International Studies.

Simon, D.F., 1992, 'Taiwan's emerging technological trajectory: creating new forms of competitive advantage', in Denis Fred Simon and Michael Y.M. Kau (eds), *Taiwan: Beyond the Economic Miracle*, Armonk, NY: M.E. Sharpe, 123–47.

Wang, J., 1995, *The Industrial Policy of Taiwan, ROC*, Taipei: Chung-Hua Institution for Economic Research.

Wang, Jiann-Chyuan and Kuen-Hung Tsai, 1995, 'Taiwan's industrial technology, policy measures and an evaluation of R&D promotion tools', *Journal of Industry Studies*, 2.

United States of America

Brooks, H., 1986, 'National science policy and technological innovation', in R. Landau and N. Rosenberg (eds), *The Positive Sum Strategy*, Washington, DC: National Academy Press.

Bush V. et al., 1960, *Science the Endless Frontier: A Report to the President on a Program for Postwar Scientific Research*, originally issued July 1945, reissued as NSF-40, Washington, DC: National Science Foundation.

Davidson, R., 1994, 'Congress in crisis once again', in G. Peele et al. (eds), *Developments in American Politics 2*, London: Macmillan.

Davis, F., 1994, 'Interest groups and policy making', in G. Peele et al. (eds), *Developments in American Politics 2*, London: Macmillan.

Ergas, H., 1987, 'The importance of technology policy', in P. Dasgupta and P. Stoneman (eds), *Economic Policy and Technological Performance*, Cambridge: Cambridge University Press, pp. 51–96.

Hart, J., 1994, 'The Presidency in the 1990s', in G. Peele et al. (eds), *Developments in American Politics 2*, London: Macmillan.

Krugman, P., 1996, *Pop Internationalism*, Cambridge, Mass.: MIT Press.

Laffin, M., 1994, 'Re-inventing the federal government', in G. Peele et al. (eds), *Developments in American Politics 2*, London: Macmillan.

Lerner, J., 1996, *The Government as Venture Capitalist: The Long Run Impact of the SBIR Program*, NBER Working Paper no. 5753, Cambridge, Mass. National Bureau of Economic Research.

Mowery, D. and N. Rosenberg, 1993, 'The US national innovation system', in R.R. Nelson (ed.), *National Innovation Systems: A Comparative Analysis*, London: Oxford University Press.

Ostry, S. and R.R. Nelson, 1995, *Techno Nationalism and Techno Globalism, Conflict and Cooperation*, Washington, DC: Brookings Institution.

Stoneman, P., 1987, *The Economic Analysis of Technology Policy*, London: Oxford University Press, Chapter 13.

8. Conclusions

8.1 INTRODUCTION

The main objectives of this chapter are, first, to provide a comparative overview of the individual country results detailed in the chapters above largely by means of typologies of relevant patterns, and second, in so doing, (a) to identify and assess the mechanisms and processes by means of which R&D priorities are set and decisions are taken in different countries; (b) to examine fund allocations criteria used; and (c) to formulate an overall assessment of the global priority setting mechanisms in national R&D policies.

A key issue addressed in this book concerns the different ways in which the allocation of funds across research priorities is made in different countries. Funds are generally allocated in a multi-level process. At the Cabinet level, a distribution across broad fields of expenditure (ministries) is decided upon. Within ministries, funds are distributed among numerous activities, one of them being R&D. Funds allocated to R&D are divided among different programmes. At the programme level certain activities are then supported and others are not.

In general, our studies indicate that in all of the countries studied explicit criteria (for example, target rates of return) are only used in decisions on whether to support a particular activity at the programme level. There, forms are filled out, claims are made and evaluated, arguments pro and con are weighed. Even there, however, the criteria used are usually quite vague and broad; they often refer to the qualities of the proposed project, the detail of the workplan, the economic or other usefulness of the prospective results, and the competence of the proposing team. As far as our research has provided insight into these matters, it seems that we find considerable convergence across Europe in the use of mechanisms and procedures. Ever more government-funded research is being carried out under contract, ever more funds are allocated using a competitive tender system, ever more technology projects are monitored during their execution and are evaluated systematically after using 'managerial criteria'. There also seems to be a tendency to rely more on externally recruited teams of experts in project selection and a general attempt to avoid open-ended support schemes with unspecified limits to their budget.

More interesting are the criteria that are used to determine the overall funding of R&D activities and its allocation across ministries and programmes. In general, however, we find that here no explicit criteria exist. The distribution of monies is the result of a political process that is guided by what we have called 'philosophies'. These philosophies encompass an analysis of the problems society faces (the concerns of government) and a view on what should be done about them (the ambitions of government). These philosophies are partly embedded in and partly determined by what we have termed 'structures', that is, the organizational entities (institutions and bureaucracies) that share the responsibility and decision making power with regard to R&D fund allocations. It is on these philosophies and structures that we concentrate in this concluding chapter.

Before doing this, however, we shall express two reservations regarding our approach. First, perceptions, philosophies and policies are continually changing and what is observed today is only a snapshot at one point in time. However, R&D policies in place may change only slowly. There is considerable momentum in existing policies and as such it may be difficult to change policy directions and instruments with any great speed. This suggests that the current position may be what it is, at least to some degree, because of historical precedence. The policies in place may therefore reflect current attitudes and concerns not all that closely. An example might be the emphasis upon government support for defence R&D. The extent of such support may well be a reflection more of the cold war than of today's less threatening climate, but because of sluggishness it will be some time before actual spending priorities and patterns fully reflect the new philosophies. Second, previous material indicates right from the outset that it is more likely that the differences which exist across countries will be differences of emphasis rather than clear bifurcations in policy stances. This is perhaps what one would expect, given the importance of history. But it is also what one would expect, given that the differences in concerns, philosophies and procedures are also matters of differences in emphasis rather than clear-cut divisions.

This chapter concentrates upon the five main case study countries: the UK, the Netherlands, Germany, France and Finland. Only cursory references are made to the non-EU countries discussed in Chapter 7. The method is primarily to look for differences and commonalities across the case study countries. An attempt is made to produce individual typologies of national concerns, philosophies, structures and policies.

The next section considers typologies that already exist in the literature. In section 8.3 a typology of national concerns is produced; in section 8.4 a typology of philosophies and structures; and then in section 8.5 a typology of policies. In section 8.6 we consider how closely one can map across countries from characteristics through mechanisms into policies. Finally, in section 8.7 some brief implications for policy design are drawn.

8.2   EXISTING TYPOLOGIES

There are a number of existing typologies in the literature. These tend to relate to the policies that are actually in place rather than the economic concerns or policy setting procedures that underlie them. The two most quoted are those attributable to Rothwell (1983) and Ergas (1987).

Rothwell studied industrial innovation policies (and the emphasis upon industrial innovation is an important one) over six countries (Canada, Japan, the Netherlands, Sweden, the UK and the US) classifying both the main policy instruments and the tactical objectives at which the policies were aimed. He argues that there were three distinct differences in approaches to technology policy across the six countries:

1. Countries differed in terms of the tools or policy instruments used. He argues that in the US and UK policy was primarily directed at creating the right climate for encouraging technical advance and that the tools used reflected this. In Canada, Japan and the Netherlands the emphasis was placed upon supply-side tools such as the provision of financial and technical assistance including the establishment of a scientific and technological infrastructure.
2. The second difference, and the one considered most important by Rothwell, lies between those countries (such as Japan) which had clear, long-term strategies towards technological change and those (such as the US) which had left technology choice largely in the hands of private companies. This is mainly an issue of targeting.
3. The third difference was at a more general level and distinguished between those countries in which state intervention was seen as a major part of a process of indicative planning (for example, France) and those where industrial innovation policy was seen as just one part of general economic policy aiming to create a favourable climate for industrial development.

Such a typology will have echoes in what is done below but one should note that:

●   this work of Rothwell referred to the position as of the early 1980s. The situation may well have changed since (for example, US technology policy has become less hands off);

- the six countries only partly overlap with our main case study countries;
- most importantly, the Rothwell work only refers to industrial innovation policies. The work thus excludes defence-related policies, policies on science and education and also the interface between science policy and technology policy.

The second typology is due to Ergas (1987). Ergas looks at technology policy encompassing both industrial innovation and military R&D but again excludes science policy expenditures. Ergas suggests the following three-fold classification:

1. Mission-orientated countries (UK, US, France) in which big science is deployed to meet big problems. Such countries are engaged in the search for international strategic leadership, concentrate on high-technology research and have expenditures on defence R&D that dominate total government expenditure on R&D.
2. Diffusion-orientated countries (Germany, Switzerland and Sweden), the specific features of which are policies designed to provide a broadly based capacity for adjusting to technological change throughout the industrial structure by the provision of innovation-related public goods, notably in the fields of education, product standards and co-operative research.
3. The third category comprises Japan alone where co-ordinated efforts have been deployed to advance national technological goals while at the same time also emphasizing a broadly based capacity to diffuse by supplying innovation-related public goods.

This typology shows a rather different emphasis to that of Rothwell, although there are parallels. The Ergas work postdated that of Rothwell by four years but of course may no longer be completely relevant. In the UK, for example, there is now a much increased emphasis on diffusion-related policy.

Rothwell (1993) also provides a classification of technology policy instruments that will be useful for further reference below. This is presented as Table 8.1.

8.3 NATIONAL CONCERNS

The five main case study countries have certain characteristics in common. They are all advanced industrial economies (although they do differ in the relative importance of industry, services and agriculture). They are also all

Table 8.1   Technology policy instruments

Policy tool	Examples
1. Public enterprise	Innovation by publicly owned industries, setting up of new industries, pioneering use of new techniques by public corporations, participation in private enterprise.
2. Scientific and technical	Research laboratories, support for research associations, learned societies, professional associations, research grants.
3. Education	General education, universities, technical education, apprenticeship schemes, continuing and further education, retraining.
4. Information	Information networks and centres, libraries, advisory and consultancy services, databases, liaison services.
5. Financial	Grants, loans, subsidies, financial sharing arrangements, provision of equipment, buildings or services, loan guarantees, export credits, etc.
6. Taxation	Company, personal, indirect and payroll taxation, tax allowances.
7. Legal and regulatory	Patents, environmental and health regulations, inspectorates, monopoly regulations.
8. Political	Planning, regional policies, honours or awards for innovation, encouragement of mergers or joint consortia, public consultation.
9. Procurement	Central or local government purchases and contracts, public corporations' R&D contracts, prototype purchases.
10. Public services	Purchases, maintenance, supervision and innovation in health services, public building, construction, transport, telecommunications.
11. Commercial	Trade agreements, tariffs, currency regulations.
12. Overseas agents	Defence sales organizations.

Source: Rothwell (1983).

open economies reliant upon trade and their own international competitiveness for their prosperity. Finland and the Netherlands are smaller than the other three and are aware of the disadvantages of this. Despite the similarities, however, each of the countries has its own particular characteristics and particular concerns based upon its current economic situation, its history and its national debates.

In this section a summary classification of these characteristic and concerns is produced. Of particular interest is how the countries differ from each other and thus some emphasis is placed upon differences as opposed to similarities.

The approach is on two levels: on the first, the general economic situation and environmental characteristics and concerns are considered: on the second, particular characteristics and concerns as regards the R&D process are detailed. There will of course be connections between these two. The summary material is presented in Tables 8.2 and 8.3, where each country is given a rating of l, m or h, representing for each issue whether it is of low, medium or high importance in the country. Where there is no entry the authors are unable to make a judgement on the basis of the evidence available. It is

Table 8.2   General economic characteristics and concerns

Characteristic	UK	NL	G	Fr	Fi
High wage	l	h	h	h	h
Role of SMEs	l	l	h	l	l
Regional autonomy	l	l	h	m	l
Poor productivity performance	h	m	m	m	m
Low growth		m		h	
International competitiveness	h	h	h	l	h
Globalization	l	h			h
Public spending reduction	h	m	l	h	h
Inflation	h	l	m	m	l
Unemployment	l	m	h	m	h/m
Level playing fields	l	h	m	l	m
Industrial structure	l	h	l	l	h
European integration	l	m	h	h	l/m
Decentralization	l	m	m	h	l
Inward investment	h	h	m	l	h

Note: In the case of characteristics, entries indicate level of importance of the characteristic; in the case of concerns, entries indicate degrees of concern, where h = high, m = medium, l = low.

Table 8.3 National R&D concerns

Concern	UK	NL	G	Fr	Fi
R&D/GDP	l	h	h	h	h
Innovativeness	h	m			
Science–technology interface	h	h	l	l	m
Output of scientists and engineers	m	h	l	l	m
Loss of R&D overseas	l	h	m	l	h
Attracting R&D	h	m	m	l	m
High government R&D spend	h	l	l	l	m
Industry R&D spend	m	h	m	m	m
Industrial R&D breakdown	l	h	l	l	m
Benefits from technology	h	h	l	l	l
Concentration of R&D	h	h	l	l	l
SMEs	h	m	h	m	m
IT	l	m	h	m	h
Defence	h	l	l	h	l

Note: Entries indicate degrees of concern, where h = high, m = medium, l = low.

necessary to state at the outset that, although the classification is based upon the case study material above, the gradings are, to some degree at least, subjective to the authors of this book.

Table 8.2 lists three general economic characteristics across which countries differ: high wage; role of SMEs; and regional autonomy. The rankings indicate in turn whether: the economy is considered (at home) as a high wage economy; whether SMEs play a large role in the production system; whether there is much regional autonomy in the economy and the political system. Of course, many other economic characteristics could have been listed but the five economies are in fact very similar in many of these other dimensions (for example, importance of trade, level of development and so on).

A number of issues that have been isolated as matters of concern in particular countries are then listed. It should be noted that the scoring of these factors is to be taken as a reflection of national concern and not as an outsider's view of the relative importance of the different factors. It should also be noted that some of these concerns are very similar, for example, international competitiveness and poor productivity growth. How different countries view the essence of a particular problem is an important part of the determination of the policies adopted to address that problem (thus, for example, poor international competitiveness could be tackled with devaluation; poor productivity performance by technology subsidies).

Going through the factors listed: poor productivity performance basically indicates a national concern with output per head of the economy; low growth indicates a concern with the rate at which the macroeconomy is growing; international competitiveness is a concern that the economy has a poor international trading performance (on the grounds of price, quality or industrial mix); globalization refers to a concern with the growing internationalization of the world economy and in particular the footloose nature of multinational companies and probably the growing trading strength of developing countries; public spending reduction refers to the emphasis that government places upon control of either total public spending or the public sector borrowing requirement; concern with inflation is self-explanatory as is concern with unemployment; level playing field refers to whether the need to match policies in other countries is an important issue; industrial structure concerns whether countries are concerned with the mix of industries within the economy (Finland is worried about its 'wooden legs' and the Netherlands about its emphasis on lower- and medium-tech industries); European integration relates to whether countries have a particular concern with promoting European integration; decentralization concerns whether there is any interest in devolving power to regions; inward investment concerns whether the country particularly emphasizes the need to provide a warm welcome to direct investment from overseas.

In addition to the above, one can identify particular issues that are considered important in particular countries but are country specific. Thus, for example, in France the policy of maintaining a strong franc in order to secure entry to the Single European Currency was a particular influence. In Germany reunification and the mentoring role for the Central and East European states is particular. Finland is special both in the impact of recent changes in the ex-Soviet Union (a main market previously) and in its recent membership of the EU.

Table 8.3 concentrates upon R&D concerns, that is, the issues relating to the R&D process in the economy that seem to be dominant in each country. Again in some cases the issues, although presented as different, are matters of emphasis. For example, whereas most countries have a particular concern with the ratio of GERD to GDP, the UK is primarily interested in the innovativeness of the economy. Such innovativeness may however derive from a high ratio of R&D to GDP.

Going through the particular concerns, the gradings relate to the following: Is the R&D/GDP ratio a particular target of policy or of particular concern? Is the economies' innovativeness (the successful exploitation of new ideas) a particularly important R&D issue? Is the science–technology interface an important concern in that there is considered to be a need to link science more closely to technology? Is there concern that the universities are produc-

ing too few scientists and engineers? Are there fears regarding the location
and loss of R&D overseas? Is the attraction of R&D facilities from overseas a
concern of policy makers? Is the government R&D spend considered to be
too high? Is industry R&D spend considered too low? Are there concerns
over the industrial breakdown of the R&D spend? Are there concerns over the
benefits that are realized in the economy from R&D? Is the concentration of
R&D in a few large firms an issue? Is IT a particular concern in the economy?
And is there a particular emphasis upon defence?

Just as with Table 8.2, there are also other concerns that are country
specific. For example, Germany places particular emphasis upon the contribution that the R&D system can make towards solving global environmental
problems whereas Finland emphasizes the need to consider the national
innovation system. In France a particular concern is the interface between
government-performed R&D and the private use of the results of that research (which will be a reflection of the importance of government-performed
research in France).

A useful way of summarizing the contents of these tables is to look at the
difference between countries in terms of concerns and ambitions.

Concerns

Government circles in all the countries we studied are acutely aware of rising
international competitive pressures. Capacity to innovate is seen by all as the
main source of competitive strength in global markets. In all countries a
certain nervousness and a sense of urgency with respect to technological and
managerial renewal can be felt. There are, however, certain differences in
emphasis across countries.

The United Kingdom is the home of a number of strong, large companies
serving global markets: those that have survived the tough macroeconomic
and industrial policies of the 1980s and have emerged 'leaner and fitter'.
However, there is also a broad range of smaller companies which have more
difficulties in exploiting opportunities provided by new technology and managerial innovations. Though the UK science base is perceived to be of a high
standard, diffusion and commercial implementation of its results needs to be
improved.

These concerns are less of an issue in Germany, where the *Mittelstand*, the
broad layer of SMEs, has traditionally shown technological prowess and been
an important growth engine of the economy. A problem is that the strongholds of German technological capacity are concentrated in areas that are
considered 'medium technology' and that are commercialized in mature industries. A further main concern in Germany is related to the need to extend
the national system of innovation to the East, to the *neue Bundesländer*. Also,

co-operation in innovation and technological development is seen as an indispensable element in the processes of further European integration and of establishing stable and beneficial relationships with neighbouring countries in Central and Eastern Europe.

In France the main concerns relate to the organization of the national system of innovation. The French system is relatively centralized and operates mainly in a top-down fashion. This type of organization is increasingly felt to be incompatible with the need to respond flexibly to market needs.

Issues of importance in the Netherlands are (a) fears that national global companies are becoming increasingly footloose, and (b) concerns about the attractiveness of the national economy in the eyes of international business. As a small and very open economy the Netherlands has to cope with conditions on international markets; more than in large countries, any attempt to hide from international competition is likely to backfire. (In Germany concerns about attractiveness for inward high-tech investment also tend to be voiced, but in France this does not seem to be an issue. The UK has no worries here as it sees itself as quite successful in attracting foreign investment in high-tech manufacturing industries.) Another Dutch concern relates to its industrial structure, with specialization in medium- and low-technology industries, and its apparent limited ability to turn new technologies into growing GDP and employment.

The main concerns in Finland, finally, are a consequence of two important developments: the economic collapse of the former Soviet Union and Finland's accession to the European Union. These developments necessitate that Finland reduce its dependence on its traditional industries (wood, pulp, paper and related engineering) and broaden its industrial base. As the Finnish economy is still recovering from the deep recession that followed the collapse of the Soviet Union there are fears that industrial restructuring and adoption of new technologies may worsen the employment situation.

The importance of global problems and environmental concerns (energy supply, global warming, the ozone hole) for policy making is difficult to judge as much 'window dressing' may be going on. However, whereas the UK and France have more of an eye for issues of the international balance of political power and military strength, the global environmental issues seem to figure more prominently on the policy agendas of Germany and the Netherlands.

The ability of the national system of innovation to be responsive to the needs of industry and of society at large are major issues of concern in all countries. However, these issues are most prominent in France and the UK, where research is predominantly mission-driven and where communication between the research establishment and parts of industry, especially SMEs, has been least developed. The traditional structure of the industrial base is a matter of concern in three of our five case study countries: Germany, the Netherlands and Finland. In these countries leading sectors such as electronics and telecommunications are underrepresented and there are fears of losing out on the benefits to growth and employment of new technologies.

Ambitions

Getting hold of one's fair share in international markets by boosting competitiveness through innovation is the universal aim of all the governments in our case studies (and of most others that we did not study in detail). Innovation is seen as the main escape route for high wage economies under attack in global markets from low wage newcomers. Increasing (downward) flexibility of wages and labour conditions as a defensive strategy in global competition is not judged to be a viable or attractive route for most European governments (except maybe in the UK, where it is still not preferred, however). There is also a general consensus that for innovation to play its role in building a competitive economy, it is necessary to make the science and technology base more responsive to the needs of industry (making scientific and technological development demand driven). Although once again there are differences in emphasis across countries.

France has recognized that its centrally organized research system no longer delivers what is needed. It is now in the middle of a slow and difficult process of devolution, of turning the system upside down, moving power and initiative to lower hierarchical (regional) levels. However, the French government feels that a degree of central guidance is indispensable and has taken it upon itself to formulate a new strategic view, to mobilize research around priorities and to reinforce public research.

Whereas France tries to reorganize itself to be more competitive, the UK has tried to deregulate itself to the same end. The UK has pursued lower taxes, lower inflation, lower public sector borrowing requirements, and more market and less state interference in the interests of private industry. The UK government now considers its task as to enable, to facilitate and to create favourable conditions, rather than to formulate strategic missions and identify national priorities. Conditions should be such that the science base is accessible to the productive sectors of the economy developing the knowledge that industry and society need. Government helps by providing information to SMEs, by promoting benchmarking, by supporting demonstrator projects, by fostering good management practice in businesses, but prefers to maintain an arm's-length relationship with the research community.

Government in Germany is quite happy with the organization of its science base and the relationship to the manufacturing sectors of the economy. As for the aims of technology policy, they tend to be related to other responsibilities and policies. There is support for the development of competence in fields of science in which German manufacturing hopes to develop world-class competence, such as aeronautics, information technology and materials research, and there is substantial attention to environmental research. R&D policy has a role in facilitating the integration of the new *Länder* in the East by developing a knowledge infrastructure, by promoting technology-based firms and by strengthening SMEs.

Policy in the Netherlands and Finland is more pragmatic and less driven than in the UK or France by any one overriding idea of direction and less driven than in Germany by a sense of international profile. Both countries closely monitor what other players on the global scene are doing and are well aware of the ongoing policy competition between members of the European Union. In these smaller countries governments are increasingly feeling the constraints of a limited science budget and are beginning to face the necessity of setting R&D priorities for the public part of the national system of innovation. These priorities are to be decided upon jointly by industry and government, taking the needs of the private, public, profit and non-profit sectors of society into account. After its period of relative isolation Finland is now pursuing rapid integration into the European and world research community by signing up to all kinds of European research organizations and EU R&D programmes. Its drive is actively to promote modernization and differentiation of the industrial structure through innovation, vocational training, extension of the knowledge base and the like. In the Netherlands one way of giving industry the lead in technological development is to resort as much as possible to generic policy instruments (such as tax reliefs and innovation credits) and matching skill supplies to the demands of the labour market. Access to support schemes has recently been made broader than it used to be (by abolishing restrictions, for example, regarding firm size or sector of activity).

8.4   NATIONAL PHILOSOPHIES AND DECISION MAKING STRUCTURES

Each of the case studies discussed the different structures, philosophies and policies in place in each country. This section first concentrates upon the philosophies that lie behind government R&D processes and policies. Philosophies are interpreted rather widely to encompass, for example, the political complexion of the government in power, the prevailing administrative culture, the role and importance of consensus, the role of consultation and the belief in the efficiency of markets.

Table 8.4 presents a number of indicators of philosophies that underlay R&D policies in the different case study countries. These indicators have

Table 8.4   National political philosophies

Philosophy	UK	NL	G	Fr	Fi
Coalition government	No	Yes	Yes	Yes	Yes
Etatism	l	m	l	h	l
Corporatist	l	m	h	l	m
Liberal	h	m	l	l	m
Consensus seeking	l	h	h	l	h
Conflictual	h	l	l	h	l
Consultation procedures					
formal	m	h	m	m	h
informal	h	l	h	h	l
with:					
industry	h	h	h	h	h
academia	m	h	h	m	h
labour	l	m	m	l	h
Belief in free market	h	h	h	m	m
Market failure legitimization	h	h	h	l	h
Reduce state involvement	h	m	l	l	m
Decentralization	l	l	h	h	l

Note: Entries indicate relative importance of the philosophy or attitude in the country concerned: h = high, m = medium, l = low.

been isolated from the case study reports. Again the importance of the different philosophies in each of the countries have been scored using the scale low (l), medium (m) or high (h).

Some further detail on the indicators in Table 8.4 is useful. The first indicator is whether the government is a coalition or not (for Germany we are concerned here with the federal government and not the *Länder* governments). Clearly, the political culture of the governments in power is also a key issue. The UK had a radical Conservative government for the 18 years through to 1997, although it now has a Labour government. The Netherlands has a socialist liberal coalition (although the political culture there has been changing, as we observed in the case study report). In Germany the federal government is a coalition of Social Democrats and Environmentalists, whereas in Finland there is a 'rainbow coalition', encompassing conservatives, social democrats and the left-wing alliance. In France a socialist government is in office under a right-wing president (cohabitation).

Like many others, we have characterized the German administrative tradition as corporatist, the British tradition as liberal and the French as étatist.

Etatism is a system where there is heavy state involvement with the economy and a benevolent elite makes many of the key economic decisions. A corporatist structure implies a system with a heavy emphasis upon consensus decision making. The liberal philosophy is one of individual responsibility and the legitimating power of free market forces with minimal government intervention. Consensus-seeking will be much more evident in corporatist structures and conflict more apparent in the liberal structure. Within all structures, however, consultation procedures are widely used. In some nations these are formalized; in others they are more informal. The main differences across countries apart from the formal/informal emphasis is the extent to which labour is involved in such consultations. A belief in free markets goes along with liberalism, using market failure to justify government intervention, and also a desire to reduce state involvement. Decentralization is primarily a characteristic of Germany and France and may well be more a reflection of the size of these countries than any other factor.

German society, both in economic and public life, is tightly organized and equipped with the organizational structures that promote interaction and consensus-seeking. The UK is characterized by much more adversarial political and industrial relationships. The management of France, including large parts of its main industrial firms, is entrusted to an élite with a common background. The Netherlands and Finland are hybrids: mainly corporatist but increasingly with liberal tendencies.

Though these descriptions are caricatures that neglect real-life subtleties, they do point to important differences between countries. In France the missions of the research establishment are formulated at the top. The research system is well equipped to handle large-scale projects in areas such as space, aerospace, nuclear energy and telecommunications. There are intimate links between government research and large industrial research establishments, especially in the arms industry, and the line between civil and military R&D is very thin. Worries about facilitating bottom-up innovations and about creating favourable conditions for innovations in SMEs are relatively low on the French policy agenda.

Decision making procedures regarding R&D are difficult to classify in a manner similar to the above. They are largely internal to the organizations involved in the R&D decision making process and thus not always clear and visible. Thus in Table 8.5 we present a summary overview of the decision making structures for R&D using a number of different indicators. In each of the case study countries the main players are the Treasury or Finance Ministry, a ministry with responsibility for industry, a ministry or ministries responsible for science and education, the defence ministry, other ministries, the Cabinet or its equivalent, Parliament, and in many cases a national advisory body for science and technology. The table first indicates (a) whether the

Table 8.5   National decision making structures

	UK	NL	G	Fr	Fi
Integrated science and education (Y/N)	N	Y	Y	Y	Y
Importance in R&D policy of:					
Parliament	l	l	h	m	l
Cabinet	m	m	m	m	m
Treasury/finance ministry	h	l	m	h	h
Science/education ministry	h	h	h	h	h
Industry ministry	h	h	h	h	h
Defence ministry	h	l	m/l	h	l
Other ministries	l	m	l	m	l
Advisory board	h	h	m	l	l
Regions in higher education	l	l	h	l	m
Foresight body (Y/N)	Y	Y	Y	Y	Y

science and education ministries are separated; (b) the relative importance in the determination of the R&D budget of the different institutions. Next, the importance of regional institutions in the university sector is indicated. Whether there are any institutions looking at the balance of priorities in the science budget (for example, Technology Foresight in the UK, the Science Council in Germany and the Consultative Committee on Government Surveys in the Netherlands) is also indicated.

As with other indicators, some countries also have special characteristics; for example, although each of the case study countries works on a frame-budgeting principle, France is different in having one budget for R&D across all ministries.

In most countries the main spending departments with a role to play in the R&D game are the ministries of science and education and of trade and industry. In the larger countries the ministry of defence is also a main spender. In the UK, unlike the other case study countries, only a small part of public money for science in universities flows through the ministry dealing with education; substantial amounts pass through a division of the Department of Trade and Industry (the Office of Science and Technology). In all countries the budgets of the spending departments are supervised by the Treasury. However, the extent to which the Treasury is involved in the actual process of policy development differs across countries. Involvement seems to be strong-est in the UK, where the Treasury both considers the justification for spending offered by ministries *ex ante* and *ex post* and monitors the effectiveness of public money spent. The UK Treasury seems to have a narrow conception of market failure and to be keen on the additionality of taxpayers' money spent on research.

Because, commonly, various ministries are involved in public spending on science and technology there is in most countries a system of co-ordination of budget setting across government. In the Netherlands, the Minister of Education, Culture and Sciences co-ordinates spending on science, whereas his or her colleague in Economic Affairs overlooks spending on technology. In the UK the Office of Science and Technology is the co-ordinating institution, and in France integration is brought about by having a common interministerial budget for science and technology. However, despite all these attempts to co-ordinate, it appears that interdepartmental co-ordination and cooperation is often limited. Usually individual ministers are fairly autonomous as far as their own budget is concerned and compete among each other for funds and for credits from parliament and public.

Remarkably, the involvement of parliament and of labour unions in the process of R&D budget setting is quite low key in all the countries we considered. In France, substantial R&D spending is supported by parliament but decisions on details of how to spend and what policy instruments to use are left to the executive. The situation is similar in the Netherlands and the UK, but in Germany parliament may be slightly more involved through a number of sub-committees.

In all countries, in matters of R&D policy, informal networks comprising industry and the government bureaucracy, especially the Ministry of Trade and Industry, are important and seem to work smoothly. However, the coverage of these networks is possibly broader in countries with a strong *Mittelstand* such as Germany and in smaller countries like Finland and the Netherlands than it is in the UK and in France. In Finland these ties even have a formal character, as both representatives of private industry and the research establishment as well as government ministers are members of the main policy making council.

A particular difference across countries is the organization of the science base. In both France and Germany the strongest parts of the science base are predominantly located in public research institutes. Authorities at the regional level are responsible for universities and polytechnics, whose main mission is education (the research side is often less well developed). Using this local knowledge base, and also the broad-based system of research organizations in different industries, German regional authorities are active in stimulating innovation in the private sector, serving the needs of local SMEs through information and consultation, diffusion policies and support of start-ups. In France, despite attempts at devolution, they have found it difficult to get this regional component of the innovation system up and running.

In the UK the main strongholds of the science base are integrated in universities. Many of the public research institutes that used to exist have been privatized. In the UK the cultural distance between the universities and industry always used to be large. Over the last couple of years the relationship between the government and the scientific establishment has become increasingly tense as the government attempts to move the science and technology base closer to private industry (in order to facilitate the diffusion of new technology to broader layers of industry, especially to SMEs) while still maintaining an arm's-length relationships between itself and the universities.

In both the Netherlands and Finland basic research is concentrated in universities and public applied research is to a large extent performed in research institutes. Because in Finland universities are the responsibility of the many regional authorities, the system there is a bit fragmented. In the Netherlands the universities deal directly with the Ministry of Education.

It is worth noting as a final comment in this section that organizational structures are usually quite inflexible and long-lived; the allocation of funds also appears to a considerable degree to be determined by historical circumstances and to show great stability over time.

8.5   POLICY ORIENTATIONS AND INSTRUMENTS

This section provides classifications of policy attitudes and instruments. The first classification is based upon the work of Rothwell (1983) and Ergas (1987), discussed above. However, whereas Ergas and Rothwell attempted to provide simple opposing classifications of countries (for instance, a country was either mission-orientated or diffusion-orientated), this is not a fair reflection of what is observed. Instead countries differ primarily in terms of the emphases apparent in their policy stances. Thus, all countries show some indications of both mission orientation and diffusion orientation. It is thus more informative to classify countries by the emphasis placed upon these orientations in their policy stances. Therefore the low, medium, high classification used above is used again here.

Table 8.6 details the nature of the policy stances using the Ergas and Rothwell classifications in each of the case study countries, based upon the material discussed above. The first policy characteristic is mission orientation and the second diffusion orientation. Ergas defines mission orientation as involving a search for international strategic leadership, with a concentration on high technology research and high expenditures on defence R&D. Diffusion orientation is reflected in policies designed to provide a broadly based capacity for adjusting to technological change throughout the industrial structure by the provision of innovation-related public goods, notably in the fields of education, product standards and co-operative research.

Table 8.6   Policy orientations and instruments: Ergas (1987) and Rothwell (1983) typologies

	UK	NL	G	Fr	Fi
Ergas typology					
Mission orientation	h	l	l	h	m
Diffusion orientation	m	h	h	l	h
Rothwell typology (civil industrial innovation)					
Creating climate	h	m	m	l	m
Provision of assistance	l	h	l	h	m
Strategic direction	m	l	l	h	m
Private choice	h	h	h	l	m
Indicative planning	l	l	l	h	l
Part of general policy	l	m	m	l	h

France and the UK are rated as having high degrees of mission orientation. This is clearly the case for France, as seen in the case study material. For the UK it primarily reflects the emphasis upon defence R&D in total government R&D spending. For Germany, the Netherlands and Finland the degree of mission orientation is clearly much less, although noticeable to some degree.

Germany and the Netherlands are rated as having high diffusion orientation. Finland is also highly rated largely because of the 'innovation system' approach that comes across throughout the case study. France is considered to have low diffusion orientation: there is evidence in the case study that creating conditions for innovation is a low policy priority. The UK, on the other hand, now places considerable emphasis upon creating 'friendly environments' and we rate this as having medium diffusion orientation (the high defence spend prevents one from giving a higher score).

Although to a degree it covers some of the same material as the Ergas classification, it is worth also exploring the policy classification proposed by Rothwell. This classification, it should be recalled, relates only to industrial innovation. The ratings here are restricted to the civil side of R&D. It contrasts policies in three dimensions: creating the right climate versus the provision of financial and technical assistance; the establishment of clear long-term strategies versus leaving technology choice largely in the hands of private companies; and intervention seen as a major part of a process of indicative planning versus industrial innovation policy seen as just one part of general economic policy aiming to create a favourable climate for industrial development. In Table 8.6 ratings are given for all six dimensions rather than contrasting each pair.

The ratings on creating the right climate are based upon a reading of the case study material. The ratings on provision of assistance are based upon the case study material plus data on the share of GERD financed by industry (netting out the high levels of government R&D defence spending in the UK and France). The ratings on strategic direction are high for France, reflecting the case study material, medium for the UK and Finland respectively and low for Germany and the Netherlands. The ratings for private choice primarily reflect a lack of strategic direction except for the UK, where the political emphasis on the free market has led to the high rating. On indicative planning, only France emphasizes this dimension. As to whether innovation policy is just one part of general policy: Finland is rated high because of its 'national system of innovation' approach; the UK low because of a policy stance that tends to consider technology policy separately from other policies; France low because of the emphasis on indicative planning; Germany in a midway position because of its policies with respect to the new *Länder*; and the Netherlands in a midway position in a comparison with the UK.

The Ergas and Rothwell typologies, although useful, are not particularly comprehensive. They do not, for example, give much insight into science policy issues. It is thus useful to supplement this by looking at policy at a lower level of aggregation and considering the importance of different policy instruments in different countries. In Table 8.1, taken from Rothwell (1983), a list of policy instruments is given. Table 8.7 is an adaptation of it. Although not fully comprehensive in terms of policy instruments, this list indicates certain policy dimensions that are not reflected in the Rothwell and Ergas typologies and on which the case study countries do differ. Scores on the importance of the different instruments in different countries using the now familiar h, l, m scale are presented in Table 8.7.

The scoring in Table 8.7 reflects a number of different pieces of evidence. As regards publicly owned enterprises, the main issue is the size of the publicly owned sector (in the UK, for example, nearly all utilities are now privatized). The scoring on public research institutes and university basic research reflects the different science structures in different countries. The scoring on tertiary education reflects international comparative data on university spending. Information spreading and grants to industry scores are based on case study material. Military R&D scores are obvious. Taxation incentive scores reflect the material in the case studies. The use of legal and regulatory instruments appears to be low in each country. Regional policy scores reflects our view of the importance of regions in R&D policy. Procurement scores reflect a view of whether procurement is used as an R&D policy instrument. The scores on attracting overseas R&D reflect a view based on both intent and success. The scores on support for SMEs is based on case study material.

Table 8.7   Policy instruments: scores

Policy		UK	NL	G	Fr	Fi
1.	Innovation by publicly owned enterprises	l	l	l	h	l
2.	Support for public research institutes	l	h	h	h	h
3.	Support for university basic research	h	m	l	l	m
4.	Support for tertiary education	m	h	h	m	h
5.	Information spreading and networks	h	h	h	l	m
6.	Grants and loans to industry (civil)	l	m	m	h	m
7.	Support for military R&D	h	l	l	h	l
8.	Taxation incentives to R&D and innovation	l	h	l	h	l
9.	Legal and regulatory encouragement	l	l	l	l	l
10.	Regional policy	l	l	h	m	l
11.	Procurement (civil only)	l	l	l	m	l
12.	Attracting overseas R&D	h	m	l	l	m
13.	Support for SMEs	h	m	h	l	m

The differences across countries that we observe in Tables 8.6 and 8.7 also extend to countries' attitudes towards European-level R&D policies, the representation of their interests in the making of such policy, and how their national policies and spending interact with European initiatives. For example, on the first of these, Germany and France have European leadership roles but would prefer R&D policy to be separated from other policies. Finland is still a very minor player. The Netherlands is particularly concerned with its effectiveness in influencing policy. The UK attitude is rather ambivalent, wishing both to maximize the national benefit but not to be wholly committed. On the relationship between national R&D policies and budgets and European R&D policies and budgets the differences may, however, be more of perception than content. The UK 'EuroPes' system and the German Gymnicher Agreement if enforced would both achieve non-additionality. Holland and Finland have no such systems (and France has perhaps a halfway house). It appears, however, that the degrees of realized additionality do not differ to any great extent across countries. Each country considers its own policies in the light of EU policies and will wherever possible attempt to not replicate.

8.6 MAPPINGS FROM CHARACTERISTICS THROUGH MECHANISMS TO POLICY PRIORITIES

At the level of R&D programmes and projects, criteria for funding usually relate to the prospective results of an activity, the probability of attaining this result and the prospective (social) benefit to be derived from this result. At the aggregate level, where broad directions of technological progress are chosen and national priorities are set, a mixture of factors play a role: incidence of market failure; the needs of industry; the needs of society at large. Decisions are taken against the background of perceived national strengths (for example, strong SMEs in Germany, a strong defence industry in France, strong transnational corporations in the UK), of specific national concerns (such as weak links between universities and SMEs in the UK, concentrated private R&D in the Netherlands, a narrow industrial base in Finland), and of perceived opportunities (more effective exploitation of a high-quality science base, integration of the eastern *Länder* in Germany, political integration in Europe). Actual choices on budget allocations that will be made by the member states of the European Union in the future cannot be predicted. However, knowing the ambitions, concerns, structures and vested interests that characterize these different countries helps in understanding the outcomes of the complex decision making processes that are going on.

Such understanding would be made clearer if there were simple mappings between policies and concerns, structures and philosophies. This section considers whether such a mapping can be undertaken from country economic and political characteristics through policy institutions and mechanisms into policy priorities and budgets. It should be stated at the outset that, although this is possible to some degree, the mapping will be of a general rather than a specific kind and it is unlikely to be unique. A unique mapping would enable direct links to be drawn, such that it would be possible to predict that a particular combination of economic and political characteristics under a particular set of policy institutions and mechanisms would always yield a particular set of policies. We do not believe this to be the case. However, the policies that are in place will be heavily influenced by the economic and political characteristics and the set of policy institutions and mechanisms. It is this link which is discussed in this section. The most informative way to proceed is to discuss the five case study countries in turn.

The United Kingdom

UK policy shows mission orientation through its emphasis on defence R&D but also diffusion orientation in its policy on civil technology. For the latter there is an emphasis upon creating the right climate and information provision with low levels of grants and loans and a reliance on private decision making and private funding of R&D (although SMEs are given special treatment). Until recently there were no special tax incentives for R&D. The process of privatization means that publicly owned enterprises and public research institutes play a small and declining role in R&D policy. Public funding of R&D has been declining and the private share has been increasing. Basic research is undertaken in universities but government commitment to tertiary education is only at a medium level. Public funding of basic research is tied more closely to user needs and national competitiveness. Overseas investors are welcomed and encouraged.

This policy stance is a clear reflection of perceived economic characteristics and concerns. The limited role of SMEs in the UK economy singles them out for special consideration. Low wages are seen as an attractor to overseas investors. The high level of concern over poor productivity and poor international competitiveness has led to a general desire to improve technological performance and to linking basic research to user needs. The latter also reflects the high level of concern over the science–technology interface and has resulted in the introduction of the Technology Foresight exercise. The desire to control inflation through control of public spending is consistent with reduced public support for R&D.

However, political attitudes have also had a major role to play. The liberal free market stance of the UK government has led it to introduce policies that emphasize the creation of a fertile environment rather than the provision of subsidies. The reduction in public expenditure on R&D is consistent with the desire to make industry take responsibility for its own activities, to fund its own research and make its own decisions. This is also consistent with the reluctance to use a strategic overview of R&D. The reluctance to use tax incentives is consistent with a desire not to distort markets.

What is much less clear in the UK case, however, is whether the institutions and structures of decision making have had a noticeable effect upon policy. The UK system is conflictual, with widespread informal consultation and only limited formal consultation procedures. Parliament plays a minor role, the spending ministries a major role, but the Treasury is dominant. Science and technology budgets are largely separated and education and science are formally separated although inter-ministerial discussion bodies do exist.

In our view the dominance of the Treasury in the UK is an important factor in the reduced government R&D budgets, the prolonged resistance against the introduction of tax incentives for R&D, the need to provide market failure justifications for policy and the growing emphasis upon efficiency and policy assessment. However, the somewhat limited nature of the consultation procedures in place may also play a role. We were struck by how the Technology

Foresight exercise and the linking of science to user needs was dominated by the desire to improve UK international competitiveness. In the Netherlands, on the other hand, their process has put greater emphasis upon social and cultural needs. German policy shows a greater emphasis upon environmental needs.

Labour plays very little role in formal or informal consultation processes. This may be why the employment consequences of innovation do not figure significantly in UK policy discussions and why unemployment *per se* is not a major policy concern (despite its high levels). Academics do, however, play a major role in consultation and are quite an effective pressure group. This may be reflected in how the science budget has managed to be largely maintained despite significant cuts elsewhere. It could also be a factor in explaining how universities have maintained their historically high profile in basic research rather than such research being undertaken in separate research institutes, as happens in other countries. The role of industry in informal consultation procedures is consistent with the shift towards 'creating the right conditions'. The role of industry pressure might also explain why the UK still has some mission orientation in its civil R&D policy, with a concentration upon aerospace.

The dominant characteristic of UK R&D spending is, however, its high defence share. There may be three reasons for this:

1.   Historically, as a major world player it was necessary for the UK to have a strong defence sector and the pattern reflected this. If this is so, the importance of defence R&D will gradually decline.
2.   Government support for defence R&D is necessary to support the UK arms industry and its exports. If this is so, it is difficult to see why the arguments on the importance of free markets should not have extended to the defence industry and why that industry should gain so much more support than the civil sector. One can only think that the military–industrial complex may have a loud voice in the corridors of power.
3.   There are significant market failures in the defence sector, for example, appropriability problems, externalities, spin-offs and so on. This is difficult to believe.

France

Of our sample, France is the mission-orientated country *par excellence*. This is reflected in: the emphasis upon defence R&D; the concentration of government civil support on high-technology projects (in space, aerospace, nuclear energy and telecommunications); the high level of R&D undertaken within the public sector; and the emphasis in government R&D on identifying directions and then taking special initiatives. In many ways the French position is very similar to the UK position in the 1970s, when the pattern of spending was very similar. Over the last 20 years, however, the UK has moved away from this pattern. France is also special in having one government-wide civil R&D budget. This is not found in any other country.

If the policies are different from those of the other countries, are there any special characteristics in the French national concerns that might lead to the different patterns in R&D policy? A growing concern with the regions is leading France to grant more regional autonomy in R&D policy (which is perhaps a reflection of the German pattern). But there seems little in the general economic characteristics and concerns that singles France out from other countries; in fact, quite the reverse, in that (a) the strong franc policy and its economic consequences, it might reasonably be argued, would suggest that France cannot continue to afford the mission-orientated approach (and there seems to be some acknowledgement of this in the French concern that in the current economic crisis the growing needs to finance salaries in government R&D establishments is crowding out other desirable spending); and (b) France is subject to the same globalization forces that are impacting on other countries but seems less concerned about them.

In terms of R&D concerns, the emphasis on R&D/GDP as a target for policy could provide a rationale for the French having special tax incentives for R&D. However, in Germany R&D is also of high concern and there are no tax incentives. But French R&D concerns are different from those in other countries in that there appears to be much less concern in general with creating conditions in industry for firms to be more innovative. There is, of course, a desire to improve technological performance in industry but that appears to be reflected more in government expenditure on private R&D.

In terms of political philosophies, France stands out as showing a high degree of étatism; a high level of informal consultation procedures, especially with industry (and, of our sample, this is the one country where the military–industrial complex seems particularly strong); and a close relationship between civil servants and industry.

The peculiar French position, where there is one unified budget for civil R&D, may be a characteristic that is due to the nature of the government mechanism. The French system encourages ministries to compete with each other. The single budget was introduced to encourage co-operation (our study suggests, however, that this has not been the result).

It seems that the French political structure, especially the emphasis on étatism, is what primarily differentiates it from our other countries and also why its policy structure is so different. The emphasis upon central decision making and a belief in the power of the state to make decisions on R&D matters is so different from that prevailing currently in the UK that it is no surprise that policies in the UK and France are now so diverse. The power of the state combined with a loud voice for industry and defence interests is consistent with the observed mission-orientated pattern of R&D spending in France. One also obtains the impression that France has a special political mission to maintain its status in the world and to stimulate French culture. This could also be a driving force behind mission orientation.

Germany

Germany differs from our other countries in a number of ways; currently it is dealing with reunification, which has led to a particular emphasis upon policies for the new *Länder* which in many ways differ from those applying to the rest of the economy; second, it has taken upon itself certain responsibilities towards European integration that other countries have not; third, its history has meant a particular stance with respect to defence and military-related technologies, with military R&D playing a much lesser role than in France or the UK. These factors are clearly important in the determination of German policy.

On a more general level, however, Germany is also different in other ways. The major responsibilities for universities rest with the *Länder* rather than the federal government (which may be compared to France and the UK, where education to secondary level is a local responsibility but tertiary education is a central government responsibility). This is consistent with basic research in Germany being undertaken in (partly) federal government-financed research institutes rather than in the universities.

In contrast to France, Germany is the country that is, according to Ergas (1987), the diffusion-orientated country *par excellence*. This is reflected in a number of ways: the importance of the Fraunhofer institutes as knowledge-diffusion instruments (it might be worth noting that the establishment of equivalent institutions was rejected in the UK in the run-up to the Technology Foresight exercise); the importance of standard-setting institutions; and the emphasis upon vocational training in the German economy. One special circumstance that may have contributed to this German stance is the relative importance of SMEs in the economy.

Until recently, Germany was considered the most successful of the European economies. Postwar it has always shown a particular aversion to inflation but recently it has experienced the unusual situation whereby it has fallen down the international R&D league table, its unemployment has risen and its growth performance has deteriorated. Thus in many ways it is now in a similar position to other European economies. On these grounds its policy stance might not therefore be expected to be much different from that in these other countries.

The key characteristics of the German political structure are: corporatism; consensus seeking; high levels of formal and informal consultation with all sectors; a belief in free markets; and a belief in decentralization. Parliament plays an important role in R&D policy (in this it is different from the other countries); the science and education ministry is probably the major R&D player; there is a Foresight body; and the Ministry of Finance plays a medium role.

Such characteristics are consistent with an R&D policy that is: diffusion orientated; leaves technology choice to private industry; plays down direct R&D assistance to industry; emphasizes the creation of favourable conditions and information spreading, considered as part of general policy. The seeking of widespread consensus is also consistent with the inclusion of environmental, that is, anti-pollution, objectives within policy (which is also seen in the Netherlands, where consensus is important, but not in the UK, where it is not).

The high level of spending on tertiary education in Germany is consistent with the *Länder* competing with each other, and the regional emphasis in German policy is also consistent with the importance of the *Länder*. This importance, combined with a desire for consensus across the board, would also be a factor in explaining why there are a large number of co-ordinating institutions with a role to play in German R&D policy.

The Netherlands

Compared to the UK, France and Germany, the Netherlands is a small economy. It is a high-wage economy with particular concerns relating to international competitiveness, its industrial structure, globalization and attracting (and re-taining) inward investment. In R&D it is particularly worried about a mediocre ratio of R&D to GDP, a loss of R&D overseas, poor industry R&D spend, the unbalanced structure of R&D, R&D concentrated in a small number of firms, and the science–technology interface. Of the five countries studied, the Nether-lands is the one where perhaps the urge to change the organization and financing of the knowledge infrastructure is greatest.

However, apart from this, there are no particular characteristics of Dutch R&D policy that especially reflect these concerns. Its policy stance is not that different from the other countries studied. There is (a) a concern in industrial innovation policy about whether to assist failing companies – the UK has long since rejected a 'lame duck' support scheme whereas France, on the other hand, is still willing to provide considerable financial support for failing firms (many of which are state owned); (b) a reasonably high level of R&D support for private industry through government grants and aid, but this is similar to France where there is perhaps a lesser sense of crisis; and relatively generous (c) tax incentives for R&D. This could be a reflection of national concerns over R&D performance and a desire to attract overseas investment, but again such incentives are apparent in France, where this not a major concern, and not in the UK, where it is.

As stated in the case study, the Netherlands' political structure shows elements of étatism as in France (for example, the role of the state in the universities and the research institutes), corporatism as in Germany (for example, the desire to obtain ex ante support for proposed policies by all those possbly affected and to achieve consensus in decision making) and liberalism as in the UK (for example, a belief in the power of free markets). Possible reflections of these three in turn would be: (a) the importance of publicly funded research institutes as opposed to universities in the science base, the high level of funding for tertiary education, and the high level of grants to industry; (b) the emphasis on cultural and ecological objectives in the Dutch Foresight exercise as opposed to the emphasis on competitiveness in the equivalent UK exercise; (c) the use of market failure rationales for government R&D intervention.

The Netherlands does differ from other countries in terms of the emphasis upon agricultural R&D, but this seems mainly to be a result of its particular industrial structure. The lesser role played by Parliament (compared to Germany) in its decision making processes does not seem to have an important impact, whereas the lesser role played by the Finance Ministry (compared to the UK) may have played some part in the introduction of tax incentives (but, as already discussed, this could be a weak argument).

What might be of particular concern, however, is the argument presented in the case studies that the desire for consensus in Dutch decision making makes for slow processes of change. This comment has also been made with respect to Germany and Finland. The UK and France do not pursue consensus. However, it is difficult to find any indicators which would show that France and the UK manage to change policies more quickly (although the move away from near market R&D was quick in the UK, the continuing high level of defence R&D may be evidence against the argument).

Finland

The last and smallest of our case study countries is Finland. Its special characteristics are its recent membership of the EU and the collapse of one of its main markets, the ex-USSR. In terms of general economic conditions it is particularly concerned by high unemployment, international competitiveness, its industrial structure, public spending and globalization. On the R&D side, its prime concerns are the ratio of R&D to GDP, the loss of R&D overseas and the potential employment impact of new technology.

As a small country it feels that it is not attractive to overseas R&D and may well lose some of its 'domestic' R&D overseas. High unemployment is a major policy concern and the recession that brought it about has put pressure on government budgets, making additional support for R&D activity infeasible (although such support has stood up better than other spending).

The particular characteristic of Finnish R&D policy which makes it stand out from the other countries is the 'national system of innovation' concept. No other country has taken a policy stance that so emphasizes innovation and the role that all policies must play in stimulating innovation in the economy. We are unable, however, to attribute this particular stance to any single aspect of the Finnish economic environment or to its political philosophies and attitudes.

The political philosophies and attitudes in Finland play down étatism but show elements of both corporatism and liberalism. Reflecting corporatism, there is an emphasis on consensus decision making, a coalition government and high levels of formal consultation activities with both industry and labour (in fact, we would say that it is the only one of our five countries in which Labour plays an extensive role in consultation). Reflecting liberalism, there is a belief in free markets and the use of market failure arguments to rationalize policy.

The role played by labour in decision making may indicate why the impact of technology upon employment is a concern in Finland but not to any major degree in any of the other countries. As a small country, defence plays almost no role in R&D.

Finnish policy shows some mission orientation in its emphasis upon nuclear R&D but otherwise the policy could be called diffusion orientated. Public research institutes play a major role as in Germany but this does not reflect the étatism found in France. It seems perhaps more to reflect the regional responsibilities for universities (although with central government finance) and thus the need for separate national basic research institutes. Tertiary education is, however, emphasized, as in Germany where universities are a wholly local responsibility. There are no tax incentives for R&D. We have suggested above that this might be a reflection of the power of the Finance Ministry (which is currently an important player in Finnish R&D policy) but the argument is weak. The share of GERD financed by industry is second only to Germany and this could reflect the belief in free markets (as well as current government financial stringency).

At this stage it is of some interest to make a few observations about the non-European economies which we discussed in the previous chapter. Briefly, we find that in Israel technology policy has been largely driven by defence-related concerns and thus is very mission orientated and defence orientated. However, an additional special characteristic, the influx of highly qualified manpower from the Soviet Union, has led to some relaxation of this orientation and a new policy stance by which government is becoming selectively involved in certain civil sectors and technologies. In Japan it has always been felt that the government was active in supporting civil technological developments and was both selective and strategic in its support programmes. The emphasis was not wholly diffusion orientated but had many such characteristics. Despite the strategic nature of the policy there is now little evidence that the Japanese government has been more successful at picking winners than other governments. Our analysis indicates that the main different characteristic of Japanese technology policy has been the success with which the Japanese have been able to favour Japanese firms at the expense of foreign firms without the use of standard (monetary) instruments. This may well reflect differences between European and Japanese cultures.

Singapore and Taiwan, as examples of Far Eastern Tiger economies, have both (until recently) shown remarkable records of output and productivity growth, largely based on catch-up. In both cases policy has been very interventionist with targeting and generous public support. In the US, however, still the leading technological nation, on the civil side technology policy is still very hands-off. The liberal free market stance has generated a non-strategic view of such policy. On the other hand, there are very large government expenditures on defence R&D and basic science. The civil use of technical advances thus produced, however, depends more upon serendipity than on policy actions.

Overview

The purpose of this section has been to explore whether there are any mappings from concerns and characteristics through mechanisms to policy instruments. There do appear to be some such relationships that can be mapped out. Such mappings are, however, rarely unique. It must also be noted that in many cases the differences between countries are often matters of emphasis rather than being discrete. Even so, we can highlight some of the tendencies that we have observed.

It appears, from the detail provided above, to be the case that a high degree of emphasis on étatism will produce a policy stance which emphasizes mission orientation (with government R&D spending biased towards defence and prestigious high-tech projects) while corporatism will lead to policies which emphasize diffusion orientation (the provision of public goods that facilitate innovative activity), and liberalism will produce policies which emphasize the creation of the correct conditions and the importance of the free market.

In addition, a search for consensus may lead to different objectives than would a conflictual decision making structure. The contrast between UK (conflictual structure) policies emphasizing competitiveness and Dutch and German (consensus structure) concerns with more ecological and cultural objectives is an example.

Similarly, a country particularly concerned with its R&D performance (such as the Netherlands) will institute policies (for example, R&D stimulation) which differ from a country concerned with its innovativeness (such as the UK with its policies emphasizing the science–technology interface).

In many ways, however, different countries are very similar in their concerns and institutions and as such their policies have many similar elements (for example, the UK and French emphases on defence R&D despite contrasting étatist and liberal political attitudes, and the common concern across most countries with market failure as a legitimating factor for policy).

8.7 SOME FINAL COMMENTS

As a final issue, we briefly address whether the material we have covered above provides any operational conclusions for policy makers. There are definite limitations to what can be concluded from the evidence presented, especially as regards national policies. However, we raise a number of particular points:

1. A realization that all countries are concerned with their technological performance and maintain a keen interest in the policies pursued by other countries. The possibilities of 'beggar thy neighbour' policy races is a very valid argument in favour of international policy co-ordination through such bodies as the EU and WTO.
2. Different countries have different concerns and objectives: this could make international policy co-ordination difficult and agreements on international policies hard to achieve.
3. Across countries, similar institutions have different levels of importance and sometimes different roles to play; thus, for example, the research orientation of UK universities compared to those in Germany considerably complicates the operation of student exchange such as the EU's Erasmus and Socrates programmes.
4. Within a given country different institutions often operate under different philosophies; thus collaboration in defence R&D across countries may entail a very different approach than collaboration in industrial R&D.
5. The different policy stances taken in different countries mean that the concept of a level playing field must be widely defined. One country may give support for R&D through tax incentives, another through

direct government spending, a third through diffusion support instruments. These are, however, generally seen as alternative instruments and one country's use of one instrument would not generally justify another country adding that instrument to an existing set of policies that are different.

6. It cannot be said that the set of policy instruments in use in any one country is more effective than those in use in another: in fact, this study could be said to confirm that the appropriate policy instruments are country specific (and dependent upon economic circumstances, political cultures and so on.)

7. Similarly, it cannot be said that the administrative structures in one country are superior to those in another: the structures themselves reflect the cultural and historical background of the economy concerned and would not necessarily transplant well into a different cultural and historical environment.

8. We can confirm that different consultation procedures lead to different policy concerns and priorities, but again there is no obvious reason for this to lead to revised structures in any of the countries we have studied.

9. Probably the most effective operational conclusion that can be drawn from this study is the need for a realization of the heterogeneity across countries in their structures, procedures and beliefs (and the heterogeneity within countries in attitudes and objectives). The realization of such heterogeneity would be a major step towards greater R&D collaboration and perhaps an important element in those national discussions in which each country looks enviously at the position of others in the international R&D league tables and attempts to devise matching policies to generate matching performance.

10. Probably the most effective strategic conclusion of our study is that there is no single (or even any) recipe for success in innovation or success in innovation policy. What suits one country may not suit another but, in addition, it is by no means clear that any country has found a magic policy bullet.

REFERENCES

Ergas, H., 1987, 'The importance of technology policy', in P. Dasgupta and P. Stoneman (eds), *Economic Policy and Technological Performance*, Cambridge: Cambridge University Press, pp.51–96.

Rothwell, R., 1983, 'The difficulties of national innovation policies', in S. Macdonald, D. McL. Lamberton and Th. Mandeville (eds), *The Trouble with Technology*, London: Frances Pinter.

Appendix

The table below provides the reader with some information on exchange rates in order to ease the comparison between different countries' expenditures on R&D. However, we note that some exchange rates might fluctuate significantly over time. The exchange rates are the average over the month of May 1999, except for the Taiwan Dollar and Israel Shekel that are averages of the daily exchange rate on the second of June.

1 Euro	1.96	Deutschmark
1 Euro	2.20	Dutch Guilder
1 Euro	5.95	Finnish Markka
1 Euro	6.56	French Franc
1 Euro	4.23	Israel Shekel
1 Euro	129.71	Japanese Yen
1 Euro	0.66	Pound Sterling
1 Euro	1.82	Singapore Dollar
1 Euro	35.60	Taiwan Dollar
1 Euro	1.06	US Dollar

Index